Construction & Evaluation (CE)

ARE 5 Mock Exam

(Architect Registration Exam)

ARE 5 Overview, Exam Prep Tips, Hotspots, Case Studies, Drag-and-Place, Solutions and Explanations

Gang Chen

ArchiteG®, Inc.
Irvine, California

Construction & Evaluation (CE) ARE 5.0 Mock Exam (Architect Registration Exam): ARE 5.0 Overview, Exam Prep Tips, Hotspots, Case Studies, Drag-and-Place, Solutions and Explanations

Copy Editor: Penny L Kortje

ArchiteG®, Inc.
http://www.ArchiteG.com

ISBN: 978-1-61265-024-1

PRINTED IN THE UNITED STATES OF AMERICA

What others are saying about *ARE Mock Exam series* & *ARE Exam Guide series* ...

"Great study guide..."
"This was a great resource supplement to my other study resources. I appreciated the mock exam questions the most, and the solutions offer an explanation as to why the answer is correct. I will definitely check out his other ARE exam resources!

UPDATE: Got my PASS Letter!"
—Sean Primeaux

"Tried everything 4 times before reading this book and PASSED!"
"I had failed this exam 4 times prior to getting this book...I had zero clue as to what I was doing wrong. I read Ballast, Kaplan and random things on the forum but for the life of me couldn't pinpoint where I was missing it until I read THIS BOOK! Gang did an excellent job...I remember ... reading Gang's book and saying Ohhhh like 4 or 5 times. I read his book several times until I became comfortable with the information. I went in on test day and it was a breeze. I remember walking out of there thinking I couldn't believe I struggled so much before. The tips in here are priceless! I strongly recommend this book..."
—hendea1

"Add this to your ARE study"
"This was a very helpful practice exam and discussion. I really appreciated the step-by-step review of the author's approach... As I studied it last before taking the test, Gang Chen's book probably made the difference for me."
—Dan Clowes ("XLine")

"Good supplemental mock exam"
"I found the mock exam to be very helpful, all of the answers are explained thoroughly and really help you understand why it is correct...Also the introduction and test taking tips are very helpful for new candidates just starting the ARE process."
—Bgrueb01

"Essential Study Tool"
"I have read the book and found it to be a great study guide for myself. Mr. Gang Chen does such a great job of helping you get into the right frame of mind for the content of the exam. Mr. Chen breaks down the points on what should be studied and how to improve your chances of a pass with his knowledge and tips for the exam.

I highly recommend this book to anyone...it is an invaluable tool in the preparation for the exam as Mr. Chen provides a vast amount of knowledge in a very clear, concise, and logical matter."
—Luke Giaccio

"Wish I had this book earlier"
"...The questions are written like the NCARB questions, with various types...check all that apply, fill in the blank, best answer, etc. The answer key helpfully describes why the correct answer is correct, and why the incorrect answers are not. Take it from my experience, at half the cost of other mock exams, this is a must buy if you want to pass..."
—Domiane Forte ("Vitruvian Duck")

"This book did exactly like the others said."
"This book did exactly like the others said. It is immensely helpful with the explanation... There are so many codes to incorporate, but Chen simplifies it into a methodical process. Bought it and just found out I passed. I would recommend."
—Dustin

"It was the reason I passed."
"This book was a huge help. I passed the AREs recently and I felt this book gave me really good explanations for each answer. It was the reason I passed."
—Amazon Customer

"Great Practice Exam"
"… For me, it was difficult to not be overwhelmed by the amount of content covered by the Exam. This Mock Exam is the perfect tool to keep you focused on the content that matters and to evaluate what you know and what you need to study. It definitely helped me pass the exam!!"
—Michael Harvey ("Harv")

"One of the best practice exams"
"Excellent study guide with study tips, general test info, and recommended study resources. Hands down one of the best practice exams that I have come across for this exam. Most importantly, the practice exam includes in depth explanations of answers. Definitely recommended."
—Taylor Cupp

"Great Supplement!!"
"This publication was very helpful in my preparation for my BS exam. It contained a mock exam, followed by the answers and brief explanations to the answers. I would recommend this as an additional study material for this exam."
—Cynthia Zorrilla-Canteros ("czcante")

"Fantastic! "
"When I first began to prepare for this exam; the number of content areas seemed overwhelming and daunting at best. However, this guide clearly dissected each content area into small management components. Of all the study guides currently available for this test - this exam not only included numerous resources (web links, you tube clips, etc..), but also the sample test was extremely helpful. The sample test incorporated a nice balance of diagrams, calculations and general concepts - this book allowed me to highlight any "weak" content areas I had prior to the real exam. In short - this is an awesome book!"
—Rachel Casey (RC)

Dedication

To my parents, Zhuixian and Yugen,
my wife, Yongqing Fang, and my daughters,
Alice, Angela, Amy, and Athena.

Disclaimer

Construction & Evaluation (CE) ARE 5.0 Mock Exam (Architect Registration Exam) provides general information about Architect Registration Examination. The book is sold with the understanding that neither the publisher nor the authors are providing legal, accounting, or other professional services. If legal, accounting, or other professional services are required, seek the assistance of a competent professional firm.

The purpose of this publication is not to reprint the content of all other available texts on the subject. You are urged to read other materials, and tailor them to fit your needs.

Great effort has been taken to make this resource as complete and accurate as possible. However, nobody is perfect and there may be typographical errors or other mistakes present. You should use this book as a general guide and not as the ultimate source on this subject. If you find any potential errors, please send an e-mail to:
info@ArchiteG.com

Construction & Evaluation (CE) ARE 5.0 Mock Exam (Architect Registration Exam) is intended to provide general, entertaining, informative, educational, and enlightening content. Neither the publisher nor the author shall be liable to anyone or any entity for any loss or damages, or alleged loss or damages, caused directly or indirectly by the content of this book.

ArchiteG®, Green Associate Exam Guide®, GA Study®, and GreenExamEducation® are registered trademarks owned by Gang Chen.

ARE®, Architect Registration Examination® are registered trademarks owned by NCARB.

If you do not wish to be bound by the above, you may return this book to the publisher for a full refund.

Legal Notice

ARE Mock Exam series by ArchiteG, Inc.

Time and effort is the most valuable asset of a candidate. How to cherish and effectively use your limited time and effort is the key of passing any exam. That is why we publish the ARE Mock Exam series to help you to study and pass the ARE exams in the shortest time possible. We have done the hard work so that you can save time and money. We do not want to make you work harder than you have to. To save your time, we use a *standard* format for all our ARE 5.0 Mock Exam books, so that you can quickly skip the *identical* information you have already read in other books of the series, and go straight to the *unique* "meat and potatoes" portion of the book.

The trick and the most difficult part of writing a good book is to turn something that is very complicated into something that is very simple. This involves researching and really understanding some very complicated materials, absorbing the information, and then writing about the topic in a way that makes it very easy to understand. To succeed at this, you need to know the materials very well. Our goal is to write books that are clear, concise, and helpful to anyone with a seventh-grade education.

Do not force yourself to memorize a lot of numbers. Read through the numbers a few times, and you should have a very good impression of them.

You need to make the judgment call: If you miss a few numbers, you can still pass the exam, but if you spend too much time drilling these numbers, you may miss out on the big pictures and fail the exam.

The existing ARE practice questions or exams by others are either way too easy or way over-killed. They do NOT match the real ARE 5.0 exams at all.

We have done very comprehensive research on the official NCARB guides, many related websites, reference materials, and other available ARE exam prep materials. We match our mock exams as close as possible to the NCARB samples and the real ARE exams instead. Some readers had failed an ARE exam two or three times before, and they eventually passed the exam with our help.

All our books include a complete set of questions and case studies. We try to mimic the real ARE exams by including the same number of questions, using a similar format, and asking the same type of questions. We also include detailed answers and explanations to our questions.

There is some extra information on ARE overviews and exam-taking tips in Chapter One. This is based on NCARB *and* other valuable sources. This is a bonus feature we included in each book because we want our readers to be able to buy our ARE mock exam books together or individually. We want you to find all necessary ARE exam information and resources at one place and through our books.

All our books are available at
http://www.GreenExamEducation.com

How to Use This Book

We suggest you read *Construction & Evaluation (CE) ARE 5.0 Mock Exam (Architect Registration Exam)* at least three times:

Read once and cover Chapter One and Two, the Appendixes, the related *free* PDF files, and other resources. Highlight the information you are not familiar with.

Read twice focusing on the highlighted information to memorize. You can repeat this process as many times as you want until you master the content of the book.

After reviewing these materials, you can take the mock exam, and then check your answers against the answers and explanations in the back, including explanations for the questions you answer correctly. You may have answered some questions correctly for the wrong reason. Highlight the information you are not familiar with.

Like the real exam, the mock exam will continue to use **multiple choice, check-all-that-apply,** and **quantitative fill-in-the-blank**. There are also three new question types: **Hotspots, case studies,** and **drag-and-place**.

Review your highlighted information, and take the mock exam again. Try to answer 100% of the questions correctly this time. Repeat the process until you can answer all the questions correctly.

CE is one of the most difficult ARE divisions because some CE questions require calculations. This book includes most if not all the information you need to do the calculations, as well as step-by-step explanations. After reading this book, you will greatly improve your ability to deal with the real ARE CE calculations, and have a great chance of passing the exam on the first try.

Take the mock exam at least two weeks before the real exam. You should definitely NOT wait until the night before the real exam to take the mock exam. If you do not do well, you will go into panic mode and NOT have enough time to review your weaknesses.

Read for the final time the night before the real exam. Review ONLY the information you highlighted, especially the questions you did not answer correctly when you took the mock exam for the first time.

This book is very light so you can easily carry it around. These features will allow you to review the graphic section whenever you have a few minutes.

The Table of Contents is very detailed so you can locate information quickly. If you are on a tight schedule you can forgo reading the book linearly and jump around to the sections you need.

All our books, including "ARE Mock Exams Series" and "LEED Exam Guides Series," are available at
GreenExamEducation.com

Check out FREE tips and info at **GeeForum.com**, you can post your questions for other users' review and responses.

Table of Contents

Chapter One

Overview of the Architect Registration Examination (ARE)

A. First Thing First: Go to the Website of your Architect Registration Board and Read all the Requirements of Obtaining an Architect License in your Jurisdiction

See the following link:
https://www.ncarb.org/get-licensed/state-licensing-boards

B. Download and Review the Latest ARE Documents at the NCARB Website

1. Important links to the FREE and official NCARB documents

NCARB launched ARE 5.0 on November 1, 2016. ARE 4.0 will continue to be available until June 30, 2018.

ARE candidates who started testing in ARE 4.0 can choose to "self-transition" to ARE 5.0. This will allow them to continue testing in the version that is most suitable for them. However, **once a candidate transitions to ARE 5.0, s/he cannot transition back to ARE 4.0**.

The current version of the Architect Registration Examination (ARE 5.0) includes six divisions:

- Practice Management (PcM)
- Project Management (PjM)
- Programming & Analysis (PA)
- Project Planning & Design (PPD)
- Project Development & Documentation (PDD)
- Construction & Evaluation (CE)

All ARE divisions continue to use **multiple choice, check-all-that-apply,** and **quantitative fill-in-the-blank**. The new exams include three new question types: **Hotspots, case studies,** and **drag-and-place**.

There is a tremendous amount of valuable information covering every step of becoming an architect available free of charge at the NCARB website:
http://www.ncarb.org/

For example, you can find guidance about architectural degree programs accredited by the National Architectural Accrediting Board (NAAB), NCARB's Architectural Experience Program (AXP) formerly known as Intern Development Program (IDP), and licensing requirements by state. These documents explain how you can qualify to take the Architect Registration Examination.

We find the official ARE 5.0 Guidelines, ARE 5.0 Handbook, and ARE 5.0 Credit Model extremely valuable. See the following link:
http://www.ncarb.org/ARE/ARE5.aspx

You should start by studying these documents.

2. **A detailed list and brief description of the FREE PDF files that you can download from NCARB**
The following is a detailed list of the FREE PDF files that you can download from NCARB. They are listed in order based on their importance.

- All **ARE 5.0** information can be found at the following links:
http://www.ncarb.org/ARE/ARE5.aspx
http://blog.ncarb.org/2016/November/ARE5-Study-Materials.aspx
- The **ARE 5.0 Credit Model** is one of the most important documents, and tells you the easiest way to pass the ARE by taking selected divisions from ARE 4.0 and ARE 5.0.

ARE5.0:	Practice Management	Project Management	Programming & Analysis	Project Planning & Design	Project Development & Documentation	Construction & Evaluation
ARE 4.0:						
Construction Documents & Services	●	●			●	●
Programming Planning & Practice	●	●	●			
Site Planning & Design			●	●		
Building Design & Construction Systems				●	●	
Structural Systems				●	●	
Building Systems				●	●	
Schematic Design				●		

Figure 1.1 The relationship between ARE 4.0 and ARE 5.0

As shown in matrix above, if you are taking both ARE 4.0 and ARE 5.0, you can pass the ARE exams by taking only five divisions in total. To complete the ARE, your goal is to select and pass exams from both versions which cover all sixteen dots in matrix above. The quickest potential options are as follows:

a. You can take the following five divisions to pass the ARE:
   ARE 4.0
   - Construction Documents & Services
   - Programming Planning & Practice
   - Site Planning & Design

   ARE 5.0
   - Project Planning & Design
   - Project Development & Documentation

OR

b. You can take the following five divisions to pass the ARE:
   ARE 4.0
   - Construction Documents & Services
   - Programming Planning & Practice

   ARE 5.0
   - Programming & Analysis
   - Project Planning & Design
   - Project Development & Documentation

- **ARE 5.0 Guidelines** includes extremely valuable information on the ARE overview, NCARB, registration (licensure), architectural education requirements, the Architectural Experience Program (AXP), establishing your eligibility to test, scheduling an exam appointment, taking the ARE, receiving your score, retaking the ARE, the exam format, scheduling, and links to other FREE NCARB PDF files. You need to read this at least twice.

- **NCARB Education Guidelines** contains information on education requirements for initial licensure and for NCARB certification, satisfying the education requirement, foreign-educated applicants, the education alternative to NCARB certification, the Education Evaluation Services for Architects (EESA), the Education Standard, and other resources.

- **Architectural Experience Program (AXP) Guidelines** includes information on AXP overview, getting started and creating your NCARB record, experience areas and tasks, documenting your experience through hours, documenting your experience through a portfolio, and the next steps. You need to read this document at least twice. The information is valuable.

  NCARB renamed the **Intern Development Program (IDP)** as **Architectural Experience Program (AXP)** in June 2016. Most of NCARB's 54-member boards have adopted the AXP as a prerequisite for initial architect licensure. Therefore, you should be familiar with the program.

The AXP application fee is $100. This fee includes one free transmittal of your Record for initial registration and keeps your Record active for the first year. After the initial year, there is an annual renewal fee required to maintain an active Record until you become registered. The cost is currently $85 each year. The fees are subject to change, and you need to check the NCARB website for the latest information.

There are two ways to meet the AXP requirements. The **first method** is **reporting hours**. Most candidates will use this method. You will need to document at least 3,740 required hours under the six different experience areas to complete the program. A minimum of 50% of your experience must be completed under the supervision of a qualified architect.

The following chart lists the hours required under the six experience areas:

Experience Area	Hours Required
Practice Management	160
Project Management	360
Programming & Analysis	260
Project Planning & Design	1,080
Project Development & Documentation	1,520
Construction & Evaluation	360
Total	**3,740**

Figure 1.2 The hours required under the six experience areas

Your experience reports will fall under one of **two experience settings**:
• **Setting A**: Work performed for an architecture firm.
• **Setting O**: Experiences performed outside an architecture firm.

You must earn at least **1,860 hours** in experience **setting A**.

Your AXP experience should be reported to NCARB at least every six months and logged within two months of completing each reporting period (the **Six-Month Rule**).

The **second method** to meet AXP requirements is to create an **AXP Portfolio**. This new method is for experienced design professionals who put their licensure on hold and allows you to prove your experience through the preparation of an online portfolio.

To complete the AXP through the **second method**, you will need to meet ALL the AXP criteria through the portfolio. In other words, you cannot complete the experience requirement through a combination of the **AXP portfolio** and **reporting hours**.

See the following link for more information on AXP:
https://www.ncarb.org/gain-axp-experience

- **Certification Guidelines** by NCARB (Skimming through this should be adequate. You should also forward a copy of this PDF file to your AXP supervisor.)

See the following link which contains resources for supervisors and mentors: http://www.ncarb.org/Experience-Through-Internships/Supervisors-and-Mentors/Resources-for-Supervisor-and-Mentors.aspx

- **ARE 5.0 Related FAQs (Frequently Asked Questions)**: Skimming through this should be adequate.

- **Your Guide to ARE 5.0** includes information on understanding the basics of ARE 5.0, new question types, taking the test, making the transition, getting ARE 5.0 done, and planning your budget. The document also contains FAQs, and links for more information. You need to read this document at least twice. The information is valuable.

- **ARE 5.0 Handbook** contains an ARE overview, detailed information for each ARE division, and ARE 5.0 references. This handbook explains what NCARB expects you to know so that you can pass the ARE exams. ARE 5.0 uses either **Understand/Apply (U/A)** or **Analyze/Evaluate (A/E)** to designate the appropriate cognitive complexity of each objective, but *avoids* the use of **"Remember,"** the lowest level of cognitive complexity (CC), or **"Create,"** the highest level of CC.

  This handbook has some sample questions for each division. The real exam is like the samples in this handbook.

  Tips:
  - ***ARE 5.0 Handbook*** *has about 180 pages. To save time, you can just read the generic information at the front and back portion of the handbook, and focus on the ARE division(s) you are currently studying for. As you progress in your testing, you can read the applicable division that you are studying for. This way, the content will always be fresh in your mind.*
  - *You need to read this document at least three times. The information is valuable.*

- **ARE 5.0 Test Specification** identifies the ARE 5.0 division structure and defines the major content areas, called **Sections**; the measurement **Objectives**; and the percentage of content coverage, called **Weightings**. This document specifies the scope and objectives of each ARE division, and the percentage of questions in each content category. You need to read this document at least twice. The information is valuable, and is the base of all ARE exam questions.

- **ARE 5.0 Prep Videos** include one short video for each division. These videos give you a very good basic introduction to each division, including sample questions and answers, and explanations. You need to watch each video at least three times. See the following link:
  http://blog.ncarb.org/2016/November/ARE5-Study-Materials.aspx

- **The Burning Question: Why do we need an ARE anyway?** (We do not want to give out a link for this document because it is too long and keeps changing. You can Google it with its title. Skimming through this document should be adequate.)

- **Defining Your Moral Compass** (You can Google it with its title plus the word "NCARB." Skimming through this document should be adequate.)

- **Rules of Conduct** is available as a FREE PDF file at:
  http://www.ncarb.org/
  (Skimming through this should be adequate.)

C. The Intern Development Program (IDP)/Architectural Experience Program (AXP)

1. What is IDP? What is AXP?

IDP is a comprehensive training program jointly developed by the National Council of Architectural Registration Boards (NCARB) and the American Institute of Architects (AIA) to ensure that interns obtain the necessary skills and knowledge to practice architecture independently. NCARB renamed the **Intern Development Program (IDP)** as **Architectural Experience Program (AXP)** in June 2016.

2. Who qualifies as an intern?

Per NCARB, if an individual meets one of the following criteria, s/he qualifies as an intern:
a. Graduates from NAAB-accredited programs
b. Architecture students who acquire acceptable training prior to graduation
c. Other qualified individuals identified by a registration board

D. Overview of the Architect Registration Examination (ARE)

1. How to qualify for the ARE?

A candidate needs to qualify for the ARE via one of NCARB's member registration boards, or one of the Canadian provincial architectural associations.

Check with your Board of Architecture for specific requirements.

For example, in California, a candidate must provide verification of a minimum of five years of education and/or architectural work experience to qualify for the ARE.

Candidates can satisfy the five-year requirement in a variety of ways:

- Provide verification of a professional degree in architecture through a program that is accredited by NAAB or CACB.

  OR
- Provide verification of at least five years of educational equivalents.

  OR
- Provide proof of work experience under the direct supervision of a licensed architect.

See the following link:
http://www.ncarb.org/ARE/Getting-Started-With-the-ARE/Ready-to-Take-the-ARE-Early.aspx

2. **How to qualify for an architect license?**
Again, each jurisdiction has its own requirements. An individual typically needs a combination of about eight years of education and experience, as well as passing scores on the ARE exams. See the following link:
http://www.ncarb.org/Reg-Board-Requirements

For example, the requirements to become a licensed architect in California are:
- Eight years of post-secondary education and/or work experience as evaluated by the Board (including at least one year of work experience under the direct supervision of an architect licensed in a U.S. jurisdiction or two years of work experience under the direct supervision of an architect registered in a Canadian province)
- Completion of the Architectural Experience Program (AXP)
- Successful completion of the Architect Registration Examination (ARE)
- Successful completion of the California Supplemental Examination (CSE)

California does NOT require an accredited degree in architecture for examination and licensure. However, many other states do.

3. **What is the purpose of ARE?**
The purpose of ARE is NOT to test a candidate's competency on every aspect of architectural practice. Its purpose is to test a candidate's competency on providing professional services to protect the health, safety, and welfare of the public. It tests candidates on the fundamental knowledge of pre-design, site design, building design, building systems, and construction documents and services.

The ARE tests a candidate's competency as a "specialist" on architectural subjects. It also tests her abilities as a "generalist" to coordinate other consultants' works.

You can download the exam content and references for each of the ARE divisions at the following links:
https://www.ncarb.org/pass-the-are/start

4. **What is NCARB's rolling clock?**
   a. Starting on January 1, 2006, a candidate MUST pass ALL ARE sections within five years. A passing score for an ARE division is only valid for five years, and a candidate has to retake this division if she has NOT passed all divisions within the five-year period.

   b. Starting on January 1, 2011, a candidate who is authorized to take ARE exams MUST take at least one division of the ARE exams within five years of the authorization.

Otherwise, the candidate MUST apply for the authorization to take ARE exams from an NCARB member board again.

These rules were created by the **NCARB's rolling clock** resolution and passed by NCARB council during the 2004 NCARB Annual Meeting.

ARE 4.0 division expiration dates per the Rolling Clock will remain the same for the transition to ARE 5.0.

5. **How to register for an ARE exam?**
   See the instructions in the new ARE guideline at the following link:
   http://www.ncarb.org/ARE/ARE5.aspx

6. **How early do I need to arrive at the test center?**
   Be at the test center at least 30 minutes BEFORE your scheduled test time, OR you may lose your exam fee.

7. **Exam format & time**
   All ARE divisions are administered and graded by computer. The time allowances are as follows:

Division	Number of Questions	Test Duration	Appointment Time
Practice Management	80	2 hr 45 min	3 hr 30 min
Project Management	95	3 hr 15 min	4 hr
Programming & Analysis	95	3 hr 15 min	4 hr
Project Planning & Design	120	4 hr 15 min	5 hr
Project Development & Documentation	120	4 hr 15 min	5 hr
Construction & Evaluation	95	3 hr 15 min	4 hr
Total Time:		21 hr	25 hr 30 min

Figure 1.3 Exam format & time

Remote proctoring will be introduced mid December 2020. After December 13, 2020, the number of questions and time allotted will change to accommodate remote proctoring:

Division	Number of Questions	Test Duration	Appointment Time
Practice Management	65	2 hr 40 min	3 hr 20 min
Project Management	75	3 hr	3 hr 40 min
Programming & Analysis	75	3 hr	3 hr 40 min
Project Planning & Design	100	4 hr 5 min	5 hr
Project Development & Documentation	100	4 hr 5 min	5 hr
Construction & Evaluation	75	3 hr	3 hr 40 min
Total Time:		19 hr 50 min	24 hr 20 min

Figure 1.4 New Exam format & time

NCARB suggests you to arrive at the test center a minimum of 30 minutes before your scheduled appointment. You can have one flexible 15-minute break for each division. That is why the appointment time is 45 minutes longer than the actual test time for each division.

Practice Management has 80 questions and NCARB allows you 2 hours and 45 minutes to complete the exam, so you should spend an average of (2x60+45)/80=165/80= 2.06 minutes on each question.

Project Management and **Programming & Analysis** as well as **Construction & Evaluation** each have 95 questions and NCARB allows you 3 hours and 15 minutes to complete each exam, so you should spend an average of (3x60+15)/80=195/95= 2.05 minutes on each question.

Project Planning & Design as well as **Project Development & Documentation** each have 120 questions and NCARB allows you 4 hours and 15 minutes to complete each exam, so you should spend an average of (4x60+15)/120=255/120= 2.13 minutes on each question.

To simplify this, we suggest you spend about 2 minutes for each question in ALL divisions.

8. **How are ARE scores reported?**
   All ARE scores are reported as Pass or Fail. ARE scores are typically posted within 7 to 10 days. See the instructions in the new ARE guideline at the following link:
   http://www.ncarb.org/ARE/ARE5.aspx

9. **Is there a fixed percentage of candidates who pass the ARE exams?**
   No, there is NOT a fixed percentage of passing or failing. If you meet the minimum competency required to practice as an architect, you pass. The passing scores are the same for all Boards of Architecture.

10. When can I retake a failed ARE division?

You can retake a failed division of the ARE 60 days after the previous attempt. You can only take the same ARE division three (3) times within any 12-month period.

11. How much time do I need to prepare for each ARE division?

Every person is different, but on average you need about 40 to 80 hours to prepare for each ARE division. You need to set a realistic study schedule and stick with it. Make sure you allow time for personal and recreational commitments. If you are working full time, my suggestion is that you allow no less than 2 weeks but NOT more than 2 months to prepare for each ARE division. You should NOT drag out the exam prep process too long and risk losing your momentum.

12. Which ARE division should I take first?

This is a matter of personal preference, and you should make the final decision.

Some people like to start with the easier divisions and pass them first. This way, they build more confidence as they study and pass each division.

Other people like to start with the more difficult divisions so that if they fail, they can keep busy studying and taking the other divisions while the clock is ticking. Before they know it, six months has passed and they can reschedule if need be.

13. ARE exam prep and test-taking tips

You can start with Construction & Evaluation (CE) because it gives a limited scope, and you may want to study building regulations and architectural history (especially famous architects and buildings that set the trends at critical turning points) before you take other divisions.

Complete mock exams and practice questions, including those provided by NCARB's practice program and this book, to hone your skills.

Form study groups and learn the exam experience of other ARE candidates. The forum at our website is a helpful resource. See the following links:
http://GreenExamEducation.com/
http://GeeForum.com/

Take the ARE exams as soon as you become eligible, since you probably still remember portions of what you learned in architectural school, especially structural and architectural history. Do not make excuses for yourself and put off the exams.

The following test-taking tips may help you:

- Pace yourself properly. You should spend about two minutes for each question on average.
- Read the questions carefully and pay attention to words like *best, could, not, always, never, seldom, may, false, except,* etc.

- For questions that you are not sure of, eliminate the obvious wrong answer and then make an educated guess. Please note that if you do NOT answer the question, you automatically lose the point. If you guess, you at least have a chance of getting it right.
- If you have no idea what the correct answer is and cannot eliminate any obvious wrong answers, then do not waste too much time on the question and just guess. Try to use the same guess answer for all of the questions you have no idea about. For example, if you choose "d" as the guess answer, then you should be consistent and use "d" whenever you have no clue. This way, you are likely have a better chance at guessing more answers correctly.
- Mark the difficult questions, answer them, and come back to review them AFTER you finish all questions. If you are still not sure, go with your first choice. Your first choice is often the best choice.
- You really need to spend time practicing to become VERY familiar with NCARB's question types. This is because ARE is a timed test, and you do NOT have time to learn about the question types during the test. If you do not know them well, you will NOT be able to finish your solution on time.
- The ARE exams test a candidate's competency to provide professional services protecting the health, safety, and welfare of the public. Do NOT waste time on aesthetic or other design elements not required by the program.

ARE exams are difficult, but if you study hard and prepare well, combined with your experience, AXP training, and/or college education, you should be able to pass all divisions and eventually be able to call yourself an architect.

14. Strategies for passing ARE exams on the first try

Passing ARE exams on the first try, like everything else, needs not only hard work, but also great strategy.

- **Find out how much you already know and what you should study**

  You goal is NOT to read all the study materials. Your goal is to pass the exam. Flip through the study materials. If you already know the information, skip these parts.

  Complete the NCARB sample questions for the ARE exam you are preparing for NOW without ANY studying. See what percentage you get right. If you get 62% right, you should be able to pass the real exam without any studying. If you get 50% right, then you just need 12% more to pass.

  This "truth-finding" exam or exercise will also help you to find out what your weakness areas are, and what to focus on.

  Look at the same questions again at the end of your exam prep, and check the differences.

  Note: *We suggest you study the sample questions in the official NCARB Study Guide first, and then other study materials, and then come back to NCARB sample questions again several days before the real ARE exam.*

Per NCARB, with the launch of the updated Architect Registration Examination (ARE) 5.0 in December 2020, the new cutting scores are based on the following information:

"How Many Questions Do I Need Correct to Pass?

Each division of the ARE measures different content knowledge areas. The difference in knowledge areas and the relative difficulty of the questions that make up that content area vary between divisions; therefore, expectations around how many questions you will need to answer correctly also changes from division to division.

- **Project Development & Documentation and Construction & Evaluation** require the lowest percentage of scored items to be answered correctly to pass. You need to answer between **57 – 62 percent** of scored items correctly on these divisions to pass.
- **Practice Management and Project Management** require a slightly higher percentage of questions to be answered correctly to pass. You need to answer between **62 – 68 percent** of scored items correctly on these divisions to pass.
- **Programming & Analysis and Project Planning & Design** require the highest percentage of questions to be answered correctly to pass. You need to answer between **65 – 71 percent** of scored items correctly on these divisions to pass."

For detailed information, see the following link:
https://www.ncarb.org/blog/what-score-do-you-need-to-pass-the-are

- **Cherish and effectively use your limited time and effort**

Let me paraphrase a story.
One time someone had a chance to talk with Napoleon. He said:
"You are such a great leader and have won so many battles, that you can use one of your soldiers to defeat ten enemy soldiers."

Napoleon responded:
"That may be true, but I always try to create opportunities where ten of my soldiers fight one enemy soldier. That is why I have won so many battles."

Whether this story is true is irrelevant. The important thing that you need to know is **how to concentrate your limited time and effort to achieve your goal. Do NOT spread yourself too thin**. This is a principle many great leaders know and use and is why great leaders can use ordinary people to achieve extraordinary goals.

Time and effort is the most valuable asset of a candidate. How to cherish and effectively use your limited time and effort is the key to passing any exam.

If you study very hard and read many books, you are probably wasting your time. You are much better off picking one or two good books, covering the major framework of your exams, and then doing two sets of mock exams to find your weaknesses. You WILL pass if you follow this advice. You may still have minor weakness, but you will have covered your major bases.

- **Do NOT stretch your exam prep process too long**
  If you do this, it will hurt instead of helping you. You may forget the information by the time you take the exam.

  Spend 40 to 80 hours for each division (a maximum of two months for the most difficult exams if you really need more time) should be enough. Once you decide on taking an exam, put in 100% of your effort and read the RIGHT materials. Allocate your time and effort on the most important materials, and you will pass.

- **Resist the temptation to read too many books and limit your time and effort to read only a few selected books or a few sections of each book in detail**
  Having all the books but not reading them, or digesting ALL the information in them will not help you. It is like someone having a garage full of foods, and not eating or eating too much of them. Neither way will help.

  You can only eat three meals a day. Similarly, you can ONLY absorb a certain amount of information during your exam prep. So, focus on the most important stuff.

  Focus on your weaknesses but still read the other info. The key is to understand, digest the materials, and retain the information.

  It is NOT how much you have read, but how much you understand, digest, and retain that counts.

  The key to passing an ARE exam, or any other exam, is to know the scope of the exam, and not to read too many books. Select one or two really good books and focus on them. Actually <u>understand</u> the content and <u>memorize</u> it. For your convenience, I have <u>underlined</u> the fundamental information that I think is very important. You definitely need to <u>memorize</u> all the information that I have underlined.

  You should try to understand the content first, and then memorize the content of the book by reading it multiple times. This is a much better way than relying on "mechanical" memory without understanding.

  When you read the materials, ALWAYS keep the following in mind:

- **Think like an architect.**
  For example, when you take the Project Development & Documentation (PDD) exam, focus on what need to know to be able to coordinate your engineer's work, or tell them what to do. You are NOT taking an exam for becoming a structural engineer; you are taking an exam to become an architect.

This criterion will help you filter out the materials that are irrelevant, and focus on the right information. You will know what to flip through, what to read line by line, and what to read multiple times.

I have said this one thousand times, and I will say it once more:
Time and effort is the most valuable asset of a candidate. How to cherish and effectively use your limited time and effort is the key to passing any exam.

15. ARE exam preparation requires short-term memory

You should understand that ARE Exam Preparation requires **Short-Term Memory**. This is especially true for the MC portion of the exam. You should schedule your time accordingly: in the early stages of your ARE exam Preparation, you should focus on understanding and an **initial** review of the material; in the late stages of your exam preparation, you should focus on memorizing the material as a **final** review.

16. Allocation of your time and scheduling

You should spend about 60% of your effort on the most important and fundamental study materials, about 30% of your effort on mock exams, and the remaining 10% on improving your weakest areas, i.e., reading and reviewing the questions that you answered incorrectly, reinforcing the portions that you have a hard time memorizing, etc.

Do NOT spend too much time looking for obscure ARE information because the NCARB will HAVE to test you on the most **common** architectural knowledge and information. At least 80% to 90% of the exam content will have to be the most common, important and fundamental knowledge. The exam writers can word their questions to be tricky or confusing, but they have to limit themselves to the important content; otherwise, their tests will NOT be legally defensible. At most, 10% of their test content can be obscure information. You only need to answer about 62% of all the questions correctly. So, if you master the common ARE knowledge (applicable to 90% of the questions) and use the guess technique for the remaining 10% of the questions on the obscure ARE content, you will do well and pass the exam.

On the other hand, if you focus on the obscure ARE knowledge, you may answer the entire 10% obscure portion of the exam correctly, but only answer half of the remaining 90% of the common ARE knowledge questions correctly, and you will fail the exam. That is why we have seen many smart people who can answer very difficult ARE questions correctly because they are able to look them up and do quality research. However, they often end up failing ARE exams because they cannot memorize the common ARE knowledge needed on the day of the exam. ARE exams are NOT an open-book exams, and you cannot look up information during the exam.

The **process of memorization** is like **filling a cup with a hole at the bottom**: You need to fill it faster than the water leaks out at the bottom, and you need to constantly fill it; otherwise, it will quickly be empty.

Once you memorize something, your brain has already started the process of forgetting it. It is natural. That is how we have enough space left in our brain to remember the really important things.

It is tough to fight against your brain's natural tendency to forget things. Acknowledging this truth and the fact that you cannot memorize everything you read, you need to focus your limited time, energy and brainpower on the most important issues.

The biggest danger for most people is that they memorize the information in the early stages of their exam preparation, but forget it before or on the day of the exam and still THINK they remember them.

Most people fail the exam NOT because they cannot answer the few "advanced" questions on the exam, but because they have read the information but can NOT recall it on the day of the exam. They spend too much time preparing for the exam, drag the preparation process on too long, seek too much information, go to too many websites, do too many practice questions and too many mock exams (one or two sets of mock exams can be good for you), and **spread themselves too thin**. They end up **missing the most important information** of the exam, and they will fail.

The ARE Mock Exam series along with the tips and methodology in each of the books will help you find and improvement your weakness areas, MEMORIZE the most important aspects of the test to pass the exam ON THE FIRST TRY.

So, if you have a lot of time to prepare for the ARE exams, you should plan your effort accordingly. You want your ARE knowledge to peak at the time of the exam, not before or after.

For example, if you have two months to prepare for a very difficult ARE exam, you may want to spend the first month focused on reading and understanding all of the study materials you can find as your **initial** review. Also during this first month, you can start memorizing after you understand the materials as long as you know you HAVE to review the materials again later to retain them. If you have memorized something once, it is easier to memorize it again later.

Next, you can spend two weeks focused on memorizing the material. You need to review the material at least three times. You can then spend one week on mock exams. The last week before the exam, focus on retaining your knowledge and reinforcing your weakest areas. Read the mistakes that you have made and think about how to avoid them during the real exam. Set aside a mock exam that you have not taken and take it seven days before test day. This will alert you to your weaknesses and provide direction for the remainder of your studies.

If you have one week to prepare for the exam, you can spend two days reading and understanding the study material, two days repeating and memorizing the material, two days on mock exams, and one day retaining the knowledge and enforcing your weakest areas.

The last one to two weeks before an exam is absolutely critical. You need to have the "do or die" mentality and be ready to study hard to pass the exam on your first try. That is how some people are able to pass an ARE exam with only one week of preparation.

17. Timing of review: the 3016 rule; memorization methods, tips, suggestions, and mnemonics

Another important strategy is to review the material in a timely manner. Some people say that the best time to review material is between 30 minutes and 16 hours (the **3016** rule) after you read it for the first time. So, if you review the material right after you read it for the first time, the review may not be helpful.

I have personally found this method extremely beneficial. The best way for me to memorize study materials is to review what I learn during the day again in the evening. This, of course, happens to fall within the timing range mentioned above.

Now that you know the **3016** rule, you may want to schedule your review accordingly. For example, you may want to read new study materials in the morning and afternoon, then after dinner do an initial review of what you learned during the day.

OR

If you are working full time, you can read new study materials in the evening or at night and then get up early the next morning to spend one or two hours on an initial review of what you learned the night before.

The initial review and memorization will make your final review and memorization much easier.

Mnemonics is a very good way for you to memorize facts and data that are otherwise very hard to memorize. It is often arbitrary or illogical but it works.

A good mnemonic can help you remember something for a long time or even a lifetime after reading it just once. Without the mnemonics, you may read the same thing many times and still not be able to memorize it.

There are a few common Mnemonics:

1) **Visual** Mnemonics: Link what you want to memorize to a visual image.
2) **Spatial** Mnemonics: link what you want to memorize to a space, and the order of things in it.
3) **Group** Mnemonics: Break up a difficult piece into several smaller and more manageable groups or sets, and memorize the sets and their order. One example is the grouping of the 10-digit phone number into three groups in the U.S. This makes the number much easier to memorize.
4) **Architectural** Mnemonics: A combination of Visual Mnemonics and Spatial Mnemonics and Group Mnemonics.

Imagine you are walking through a building several times, along the same path. You should be able to remember the order of each room. You can then break up the information that you want to remember and link them to several images, and then imagine you hang the images on walls of various rooms. You should be able to easily recall each group in an orderly manner by imagining you are walking through the building again on the same path, and looking at the images hanging on walls of each room. When you look at the images on the wall, you can easily recall the related information.

You can use your home, office or another building that you are familiar with to build an Architectural Mnemonics to help you to organize the things you need to memorize.

5) **Association** Mnemonics: You can associate what you want to memorize with a sentence, a similarly pronounced word, or a place you are familiar with, etc.
6) **Emotion** Mnemonics: Use emotion to fix an image in your memory.
7) **First Letter** Mnemonics: You can use the first letter of what you want to memorize to construct a sentence or acronym. For example, "**Roy G. Biv**" can be used to memorize the order of the 7 colors of the rainbow, it is composed of the first letter of each primary color.

You can use **Association** Mnemonics and memorize them as all the plumbing fixtures for a typical home, PLUS Urinal.

OR

You can use "Water S K U L" (**First Letter** Mnemonics selected from website below) to memorize them:

Water Closets
Shower
Kitchen Sinks
Urinal
Lavatory

18. The importance of good and effective study methods

There is a saying: Give a man a fish, feed him for a day. Teach a man to fish, feed him for a lifetime. I think there is some truth to this. Similarly, it is better to teach someone HOW to study than just give him good study materials. In this book, I give you good study materials to save you time, but more importantly, I want to teach you effective study methods so that you can not only study and pass ARE exams, but also so that you will benefit throughout the rest of your life for anything else you need to study or achieve. For example, I give you samples of mnemonics, but I also teach you the more important thing: HOW to make mnemonics.

Often in the same class, all the students study almost the SAME materials, but there are some students that always manage to stay at the top of the class and get good grades on exams. Why? One very important factor is they have good study methods.

Hard work is important, but it needs to be combined with effective study methods. I think people need to work hard AND work SMART to be successful at their work, career, or anything else they are pursuing.

19. The importance of repetition: read this book at least three times

Repetition is one of the most important tips for learning. That is why I have listed it under a separate title. For example, you should treat this book as part of the core study materials for your ARE exams and you need to read this book at least three times to get all of its benefits:

1) The first time you read it, it is new information. You should focus on understanding and digesting the materials, and also do an initial review with the **3016** rule.
2) The second time you read it, focus on reading the parts I have already highlighted AND you have highlighted (the important parts and the weakest parts for you).
3) The third time, focus on memorizing the information.

Remember the analogy of the memorization process as **filling a cup with a hole on the bottom**?
Do NOT stop reading this book until you pass the real exam.

20. The importance of a routine

A routine is very important for studying. You should try to set up a routine that works for you. First, look at how much time you have to prepare for the exam, and then adjust your current routine to include exam preparation. Once you set up the routine, stick with it.

For example, you can spend from 8:00 a.m. to 12:00 noon, and 1:00 p.m. to 5:00 p.m. on studying new materials, and 7:00 p.m. to 10:00 p.m. to do an initial review of what you learned during the daytime. Then, switch your study content to mock exams, memorization and retention when it gets close to the exam date. This way, you have 11 hours for exam preparation everyday. You can probably pass an ARE exam in one week with this method. Just keep repeating it as a way to retain the architectural knowledge.

OR
You can spend 7:00 p.m. to 10:00 p.m. on studying new materials, and 6:00 a.m. to 7:00 a.m. to do an initial review of what you learned the evening before. This way, you have four hours for exam preparation every day. You can probably pass an ARE exam in two weeks with this preparation schedule.

A routine can help you to memorize important information because it makes it easier for you to concentrate and work with your body clock.

Do NOT become panicked and change your routine as the exam date gets closer. It will not help to change your routine and pull all-nighters right before the exam. In fact, if you pull an all-nighter the night before the exam, you may do much worse than you would have done if you kept your routine.

All-nighters or staying up late are not effective. For example, if you break your routine and stay up one-hour late, you will feel tired the next day. You may even have to sleep a few more hours the next day, adversely affecting your study regimen.

21. The importance of short, frequent breaks and physical exercise

Short, frequent breaks and physical exercise are VERY important for you, especially when you are spending a lot of time studying. They help relax your body and mind, making it much easier for you to concentrate when you study. They make you more efficient.

Take a five-minute break, such as a walk, at least once every one to two hours. Do at least 30 minutes of physical exercise every day.

If you feel tired and cannot concentrate, stop, go outside, and take a five-minute walk. You will feel much better when you come back.

You need your body and brain to work well to be effective with your studying. Take good care of them. You need them to be well-maintained and in excellent condition. You need to be able to count on them when you need them.

If you do not feel like studying, maybe you can start a little bit on your studies. Just casually read a few pages. Very soon, your body and mind will warm up and you will get into study mode.

Find a room where you will NOT be disturbed when you study. A good study environment is essential for concentration.

22. A strong vision and a clear goal

You need to have a strong vision and a clear goal: to <u>master</u> the architectural knowledge and <u>become an architect in the shortest time</u>. This is your number one priority. You need to master the architectural knowledge BEFORE you do sample questions or mock exams, except "truth-finding" exam or exercise at the very beginning of your exam prep. It will make the process much easier. Everything we discuss is to help you achieve this goal.

As I have mentioned on many occasions, and I say it one more time here because it is so important:

It is how much architectural knowledge and information you can <u>understand, digest, memorize</u>, and firmly retain that matters, not how many books you read or how many sample tests you have taken. The books and sample tests will NOT help you if you cannot understand, digest, memorize, and retain the important information for the ARE exam. Cherish your limited <u>time and effort</u> and focus on the most <u>important</u> information.

23. Codes and standards used in this book

We use the following codes and standards:

American Institute of Architects, Contract Documents, Washington, DC; ADA Standards for Accessible Design, ADA; Various International Codes by ICC. See Appendixes for more information.

24. Where can I find study materials on architectural history?

Every ARE exam may have a few questions related to architectural history. The following are some helpful links to FREE study materials on the topic:

http://issuu.com/motimar/docs/history_synopsis?viewMode=magazine

Chapter Two

Construction & Evaluation (CE) Division

A. General Information

1. Exam content

The CE division of the ARE has 95 questions which cover the following different areas.

Sections	Target Percentage	Expected Number of Items
Section 1: Preconstruction Activities	17-23%	16-22
Section 2: Construction Observation	32-38%	30-36
Section 3: Administrative Procedures & Protocols	32-38%	30-36
Section 4: Project Closeout & Evaluation	7-13%	6-12

Figure 2.1 Exam Content

Note:
After December 13, 2020, the number of questions will be reduced. See Figure 1.4.

The exam content can be further broken down as follows:
Section 1: Preconstruction Activities (17-23%)

- Interpret the architect's roles and responsibilities during preconstruction, based on delivery method (U/A)
- Analyze criteria for selecting contractors (A/E)
- Analyze aspects of the contract or design to adjust project costs (A/E)

Section 2: Construction Observation (32-38%)

- Evaluate the architect's role during construction activities (A/E)
- Evaluate construction conformance with contract documents, codes, regulations, and sustainability requirements (A/E)
- Determine construction progress (U/A)

Section 3: Administrative Procedures & Protocols (32-38%)

- Determine appropriate additional information to supplement contract documents (U/A)
- Evaluate submittals including shop drawings, samples, mock-ups, product data, and test results (A/E)
- Evaluate the contractor's application for payment (A/E)
- Evaluate responses to non-conformance with contract documents (A/E)

Section 4: Project Closeout & Evaluation (7-13%)

- Apply procedural concepts to complete close-out activities (U/A)
- Evaluate building design and performance (A/E)

2. Official exam guide and reference index for the Construction & Evaluation (CE) division

NCARB published the exam guides for all ARE 5.0 division together as *ARE 5.0 Handbook.*

You need to read the official exam guide for the PPD division at least three times and become very familiar with it. The real exam is VERY similar to the sample questions in the handbook.

You can download the official *ARE 5.0 Handbook* at the following link:
https://www.ncarb.org/sites/default/files/ARE5-Handbook.pdf

Note: *We suggest you study the official ARE 5.0 Handbook first, and then other study materials, and then come back to Handbook again several days before the real ARE exam.*

B. The Most Important Documents/Publications for Construction & Evaluation (CE) Division of the ARE Exam

1. Official NCARB list of references for the Construction & Evaluation (CE) division with our comments and suggestions

You can find the NCARB list of references for this division in the Appendixes of this book and the *ARE 5.0 Handbook.*

Note:

While many of the MC questions in the real ARE exam ***focus on design concepts****, there are* ***some questions requiring calculations****.*

In the ARE exams, it may be a good idea to skip any calculation question that requires over 2 minutes of your time; just pick a guess answer, mark it, and come back to calculate it at the end. This way, you have more time to read and answer other easier questions correctly.

A calculation question that takes 20 minutes to answer will gain the same number of points as a simple question that ONLY takes 2 minutes.

If you spend 20 minutes on a calculation question earlier, you risk losing the time to read and answer ten other easier questions, which could result in a loss of ten points instead of one.

The following is the NCARB list of top references for this division. For a longer list of relevant reference materials, please see the reference matrix at the end of this book.

Publications

The Architect's Handbook of Professional Practice
The American Institute of Architects
John Wiley & Sons, 14th edition (2008) and 15th edition (2014)

CSI MasterFormat
The Construction Specifications Institute, 2018 edition

AIA Contract Documents

Conventional Family

A101-2017
Standard Form of Agreement Between Owner and Contractor where the basis of payment is a Stipulated Sum

A201-2017
General Conditions of the Contract for Construction (This is a very important AIA document, read it at least 3 times.)

A305-1986
Contractor's Qualification Statement

A701-2018
Instructions to Bidders

B101-2017
Standard Form of Agreement Between Owner and Architect (This is a very important AIA document, read it at least 3 times.)

C401-2017
Standard Form of Agreement Between Architect and Consultant

G701-2017
Change Order (This is a very important AIA document, read it at least twice.)

G702-1992
Application and Certificate for Payment (This is a very important AIA document, read it at least twice.)

G703-1992
Continuation Sheet

G704-2017
Certificate of Substantial Completion (This is a very important AIA document, read it at least twice.)

The following are some extra study materials that are useful if you have some additional time and want to learn more. If you are tight on time, you can simply look through them and focus on the sections that cover your weaknesses:

2. **Construction Specifications Institute (CSI) MasterFormat & *Building Construction***
   Become familiar with the new 6-digit CSI Construction Specifications Institute (CSI) MasterFormat as there may be a few questions based on this publication. Make sure you know which items/materials belong to which CSI MasterFormat specification section, and memorize the major section names and related numbers. For example, Division 9 is Finishes, and Division 5 is Metal, etc. Another one of my books, *Building Construction*, has detailed discussions on CSI MasterFormat specification sections.

Mnemonics for the 2004 CSI MasterFormat

The following is a good mnemonic, which relates to the 2004 CSI MasterFormat division names. Bold font signals the gaps in the numbering sequence.

This tool can save you lots of time: if you can remember the four sentences below, you can easily memorize the order of the 2004 CSI MasterFormat divisions. The number sequencing is a bit more difficult, but can be mastered if you remember the five bold words and numbers that are not sequential. Memorizing this material will not only help you in several divisions of the ARE, but also in real architectural practice

Mnemonics (pay attention to the underlined letters):
Good students can memorize material when teachers order.
F students earn F's simply 'cause **forgetting** principles have **an** effect. (21 and 25)
C students **end** everyday understanding things without memorizing. (31)
Please make professional pollution prevention inventions **everyday**. (40 and 48)

1-Good.................................General Requirements
2-Students............................. (Site) now Existing Conditions
3-Can....................................Concrete
4-Memorize...........................Masonry
5-Material ....Metals
6-When.................................Woods and Plastics
7-Teachers.............................Thermal and Moisture
8-Order..................................Openings

9-F..Finishes
10-Students........................... Specialties
11-Earn..................................Equipment
12-F's.....................................Furnishings

13-Simply............................... Special Construction
14-'Cause............................... Conveying
21-Forgetting .... Fire
22-Principles.......................... Plumbing
23-Have................................ HVAC
25-An.................................... Automation
26-Effect............................... Electric

27-C...................................... Communication
28-Students............................ Safety & Security
31-End.................................. Earthwork
32-Everyday........................... Exterior
33-Understanding .... Utilities
34-Things.............................. Transportation
35-Without Memorizing........ Waterways and Marine

40-Please............................... Process Integration
41-Make................................ Material Processing and Handling Equipment
42-Professional..................... Process Heating, Cooling, and Drying Equipment
43-Pollution........................... Process Gas and Liquid Handling, Purification and Storage Equipment
44-Prevention......................... Pollution Control Equipment
45-Inventions......................... Industry-Specific Manufacturing Equipment
48-Everyday......................... Electrical Power Generation

Note:
There are 49 CSI divisions. The "missing" divisions are those “reserved for future expansion” by CSI. They are intentionally omitted from the list.

Chapter Three

ARE Mock Exam for Construction & Evaluation (CE) Division

A. Multiple-Choice (MC)

1. The occupant load factors are calculated based on
   a. the number of fixed seating
   b. the net floor area
   c. the gross floor area
   d. the building official's final decision
   e. the number of fixed seating or the gross floor area, but subject to the official's final decision
   f. the number of fixed seating, the net floor area, or the gross floor area, but subject to the official's final decision

2. For a residential remodel project, the framing and interior drywalls of a bathroom are not installed flat enough, and the mirror cannot be installed properly. The framing and interior drywalls have to be demolished and re-installed. The architect does not notice this defect in her routine job site visit. Who should pay for the extra costs?
   a. The general contractor should submit a change order to the owner to cover the extra costs.
   b. The general contractor or his subcontractors should pay for the extra costs.
   c. The architect since she does not notice this defect in her routine job site visit.
   d. The architect's professional liabilities insurance company since the architect does not notice this defect in her routine job site visit.

3. A contractor is building a shop building. The civil and plumbing site and floor plans only show the water line up to 5' (1524) outside of the building, and does not show the connecting water line from water main to within 5' of the building. There is a general note on the cover sheet of the plans instructing the contractors to construct a complete water supply system, including any connection to the water main. The contractor did not give the complete set of the plans to the plumbing subcontractor during the bidding, and the plumbing subcontractor has not included the cost for the connecting water line from water main to within 5' (1524) of the building in his bid based on the civil and plumbing site and floor plans. Who should pay for the extra costs?
   a. The general contractor since a general note on the cover sheet covers this item.
   b. The owner since the plumbing subcontractor has not included the cost for this item in the original bid.
   c. The civil engineer or the plumbing engineer since they have not properly shown this item on the civil and plumbing site and floor plans.

   d. The architect or her insurance company since she is responsible for coordinating the civil engineer or the plumbing engineer's work, and should have caught this conflict in the plans.

4. An architect is specifying the stainless steel urinals for the restrooms in a public park. She has a hard time finding the stainless steel urinals that meet both the ADA and the local building codes. What should she do?
   a. Try to talk with the building officials to gain an exception to the ADA and the local building codes based on economic hardship.
   b. Continue to look for the stainless steel urinals that meet the requirements.
   c. Use a stainless steel handicap toilet instead of the stainless steel urinals to meet the requirements and provide equal access.
   d. Install the stainless steel urinals installed in other public restrooms in the same geographic area.

5. An architect has designed the interior space for an office per codes. After the project is completed, the client thinks the counters are too low, and has his employees raise the height of the counters. The architect notices the revised counter heights do *not* comply with the building codes. She advises the client the revision does not comply with codes and should be corrected. The client refuses to accept the architect's advice and says his decision is final since he is paying the bills. What should she do?
   a. Do nothing since this is the client's property.
   b. Document the incident in a written report and update her project files.
   c. Document the incident in a written report and fax the report to the client.
   d. Report the incident to the building officials.

6. The right to use property of another without possessing it is called ____________.

7. The restaurant owner informs the architect she wants to delete two toilet stalls to reduce the restroom area and increase the more profitable dining area. Prior to deleting the toilet stalls, the architect should check the requirements in:
   a. the current local building codes and plumbing codes
   b. the latest IPC
   c. the latest IBC
   d. the latest IPC and IBC

8. A public school district is trying to build a new elementary school in a neighborhood. Several of the residents at the site are not willing to sell their properties. The school district can use _________________ to purchase the properties at fair market value without the residents' consent.

9. A mixed-use building is $2 million over the budget, what choices does the owner have? **Check the four that apply.**
   a. Reduce the project scope.
   b. Increase the budget by $2 million.
   c. Ask the architect to revise the design without paying extra fees.

d. Negotiate with the contractor to reduce the construction cost.
e. Bill the architect's insurance company $2 million.
f. Back charge the architect $2 million.

10. Which of the following statement is not true regarding substitution? **Check the three that apply.**
   a. A substitution is acceptable if it is cheaper than the specified product.
   b. A substitution can be acceptable if a specified product is not available.
   c. A substitution can be verbally approved by the owner and the architect.
   d. A substitution is acceptable if it is from a local manufacturer.
   e. A substitution cannot be submitted as part of the shop drawings without any special notation.
   f. A substitution must have equal or better quality and performance as the specified product.

11. Which of the following should be part of the written architectural service contract? **Check the two that apply.**
   a. Terms regarding reimbursable expenses
   b. Terms and conditions regarding additional services
   c. Professional liability insurance
   d. Breakdown of consultants' fee
   e. Type of the construction
   f. Schematic design fee

12. Which of the following cannot divert construction waste from reduce landfill? **Check the two that apply.**
   a. Using one dumpster at the site
   b. Using three dumpsters at the site: trash, plants and recycle
   c. Requiring all packaging materials at the site to be recycled
   d. Setting up a construction waste management plan
   e. Implementing the policy of "reduce, reuse and recycle" in the architect's office
   f. Seeking LEED certification for the building

13. In contract law, the doctrine that a contract can only impose obligations or confer rights to parties to the contract is called _________________.

14. Which of the followings can help assure quality control of the construction documents? **Check the three that apply.**
   a. Allowing adequate time for preparing the construction documents before bidding
   b. Having the owner review the plans
   c. Having the contractor review the plans
   d. Having a review meeting with the building officials
   e. Using checklists
   f. Using critical path method

15. After a contract is awarded, the contractor notices that he has the site plans, but does not have the legal description of the site and the owner's interest in the site. The contractor requests the legal description from the owner. Which of the following is correct?
    a. The owner can refuse to provide the information since it is private.
    b. The contractor shall go to the county recorder's office and obtain the information since the owner is not obliged to provide it.
    c. The owner shall provide the information to the contractor.
    d. The contractor shall request the information from the civil engineer, not the owner.

16. After a project is about 50% completed, the contractor hires a new superintendent. The superintendent cannot find evidence of owner's financial arrangement for payment for the project in the contractor's project files, and she requested the owner to provide a copy of the evidence. Which of the following is correct?
    a. The owner can refuse to provide the information.
    b. The owner has to provide the information.
    c. The superintendent should ask the owner's accountant for the information.
    d. The superintendent should ask the owner's lender for the information.

17. Matters regarding the rights and responsibilities of owner, contractor and the architect to the contract regarding a specific project are best handled in:
    a. Specifications
    b. General condition
    c. Supplemental condition
    d. General notes on the cover sheet of the construction drawings

18. Who should pay for the temporary power during the construction according to A201, General Conditions of the Contract for Construction?
    a. The power company
    b. The electrical subcontractor
    c. The contractor
    d. The contractor should pay for it and then get reimbursed by the owner

19. During the demolishing process of a remodel project, the contractor discovers mold after removing the gypsum board. What is the proper action for the contractor?
    a. Stop the work at the affected area, and notify the building official and the owner in writing immediately.
    b. Stop the work at the affected area, and notify the building official and the architect in writing immediately.
    c. Stop the work at the affected area, and notify the owner and the architect in writing immediately.
    d. Stop the work at the affected area, notify the owner in writing, and locate a mold specialist to solve the problem.

20. During plan check, the building official instructs the architect to add an equipment screen wall for the rooftop units (RTU) for HVAC equipment only if they are visible from two adjacent major streets by the building inspector after the RTU is installed. What is the best way for the architect to handle this in the bidding process?
    a. Ask the contractors to include the equipment screen wall as an allowance in the bid
    b. Ask the contractors to include the equipment screen wall as an alternate bid
    c. Ask the contractors to include the equipment screen wall as a change order
    d. Ask the contractors to include the equipment screen wall as an order for a minor change in the Work

21. Which of the contractor compensation methods is best for the owner if the project scope or timing is unknown, and high quality is paramount?
    a. Stipulated-sum contracts
    b. Time and materials contracts
    c. Unit-price contracts
    d. Cost-plus-fee contracts

22. During construction, the inspector requested some field tests not included in the contract documents. Who should pay for these tests?
    a. The architect since she should include all the tests required by the codes and building officials.
    b. The architect's error and omission insurance since the architect should include all the tests required by the codes and building officials.
    c. The contractor
    d. The owner

23. During construction, the owner and the contractor have a different opinion of the paint quality, what is the first step to resolve this dispute?
    a. Seeking the IDM's interpretation
    b. Mediation
    c. Litigation
    d. Arbitration

24. For a project with $10,000 allowance for a transformer, if the owner chooses to install the transformer and the actual cost of the transformer is $9,500, the contractor shall bill the owner:
    a. $9,500
    b. $10,000
    c. $9,500 plus the contractor's overhead and profit
    d. $9,500 plus the installation cost and the contractor's overhead and profit
    e. $10,000 plus the installation cost and the contractor's overhead and profit
    f. $10,000 plus the contractor's overhead and profit but not more than 15%

25. At a field visit, the architect notices the window flashings have not been installed per the contract documents. What should she do?
    a. Reject the window flashings work.
    b. Stop the work.
    c. Inform the owner and ask the owner to stop the work.
    d. Report the incident to building official.

26. Which of the following is true about the list of items to be corrected by the contractor?
    a. It should be prepared by the contractor.
    b. It should be prepared by the architect.
    c. It should be prepared by the architect, issued to the contractor, and cc the owner.
    d. It should be prepared by the architect after a joint punch walk with the contractor and the owner.

27. Which of the following construction sequence is most likely to be productive?
    a. Footing, concrete slab, framing, drywall, carpet, roofing, and painting
    b. Footing, framing, concrete slab, drywall, carpet, roofing, and painting
    c. Footing, concrete slab, framing, roofing, drywall, painting, and carpet
    d. Footing, concrete slab, framing, carpet, painting, roofing, and drywall

28. The electrical subcontractor turned off the main electrical switch for the entire floor to connect the light fixtures to the panel. The framing subcontractor did not know this. He wanted to work on the interior framing of the floor, and turned on the main electrical switch. The electrical subcontractor is killed instantly. Who is responsible to the building owner for the electrical subcontractor's death?
    a. The framing contractor
    b. The contractor
    c. The framing contractor and the contractor jointly
    d. The insurance company of the electrical contractor

29. Who should prepare the change orders?
    a. The architect
    b. The contractor
    c. The owner
    d. The superintendent

30. Which of the following inspections does an architect normally conduct?
    a. Inspection before pouring concrete for the footing
    b. Inspection before roofing
    c. Inspection before the installation of gypsum boards
    d. Inspection to determine the date of substantial completion

31. Which of the following regarding change orders and construction change directives is true? **Check the three that apply.**
    a. Change orders are for major changes, and construction change directives are for minor changes.
    b. Both change orders and construction change directives have to be signed by the owner, contractor, and the architect.
    c. When the contractor agrees, and signs a construction change directive, it should be effective immediately and should be recorded as a change order.
    d. Change orders have to be signed by the owner, contractor, and the architect.
    e. Construction change directives have to be signed by the owner, contractor, and the architect.
    f. Construction change directives have to be signed by the owner and the architect.

32. When modifications to the scope of the contract occur, which of the following is true?
    a. The modifications have to be submitted to the surety for approval.
    b. The modifications do not need to be submitted to the surety for approval.
    c. The modifications have to be submitted to the owner's insurance company for approval.
    d. The modifications do not need to be submitted to the owner's insurance company for approval.

33. Which of the following regarding boiler and machinery insurance is correct? **Check the two that apply.**
    a. It is the owner's responsibility.
    b. It is the contractor's responsibility.
    c. If there is no boiler and machinery insurance, the owner may need to pay for any related damages.
    d. If there is no boiler and machinery insurance, the contractor may need to pay for any related damages.

34. Which of the following is true regarding a waiver of subrogation in property insurance? **Check the two that apply.**
    a. It should be requested before any loss occurs.
    b. It can be requested after a loss occurs.
    c. The contractor has to disclose the subrogation provision to the insurer before purchasing the property insurance.
    d. The owner has to disclose the subrogation provision to the insurer before purchasing the property insurance.

35. When the slope of a sidewalk is more than ______% in the direction of travel, the sidewalk becomes a ramp.

36. When you are working on the construction drawings for a wood stud building, you should dimension to the __________ of the stud.

37. The five classical orders are ____.

38. Which of the following is a floor special purpose outlet?

a.

b.

c.  F

d.

Figure 3.1 Which one is a floor special purpose outlet?

39. A developer is building a shopping center. All buildings in this shopping center are one-story. The occupancy group for these building is Group ________. The most cost-effective way to design and build these building is to qualify all of them as one "building," Type _______ construction, fully sprinklered, with ___________ feet wide public ways or yards around all sides of the shopping center so that the area of all these buildings shall not be limited.

40. When the occupant load for a room is 780, the minimum number of exit(s) required for this room is ___________.

41. Which of the following are project delivery methods? **Check the four that apply.**
   a. BIM
   b. Design-Bid-Build
   c. Construction Administration
   d. Construction Management
   e. Design-Build
   f. Integrated Project Delivery

42. Which of the following project delivery methods works best for a complicated airport project that needs to continue operating during construction?
   a. BIM
   b. Design-Bid-Build
   c. Construction Administration
   d. Construction Management
   e. Design-Build

43. If quality is the most important variable for a project, which of the following is the preferred project delivery method?
   a. Cost plus fixed fee
   b. Bridging
   c. Negotiated select team
   d. Design-Build

44. In a concrete slump test, if a shear type of slump is achieved, which of the following is true?
    a. A fresh sample should be taken and the test repeated.
    b. The test is acceptable.
    c. The test is acceptable if the shear type of slump is between 2" (51) and 5" (127).
    d. The concrete is too dry.

45. If the owner hires a design architect and a production architect, which of the following is she likely to use?
    a. Design-Bid-Build
    b. Cost plus fixed fee
    c. Bridging
    d. Negotiated select team
    e. Design-Build
    f. BIM

46. Which of the following is likely to allow the owner to know the project cost at earliest stage of the project?
    a. Design-Bid-Build
    b. Cost plus fixed fee
    c. Negotiated select team
    d. Standard design-build
    e. BIM
    f. CM-agent

47. The actual dimension of a 2x4 wood stud is ____________.

48. After moving into portion of a building, the owner notices a small crack on the bathroom mirror. This crack was not on the punch (deficiency) list reviewed by the architect. What is a proper determination for the architect?
    a. The owner should be responsible for the crack since he has moved in, and the cause of the crack cannot be established.
    b. The contractor should replace the mirror even though the crack was not on the punch (deficiency) list reviewed by the architect.
    c. The architect should pay for the cost to replace the mirror because she missed the crack on the original the punch (deficiency) list walk-through.
    d. The architect's insurance company should pay for the cost to replace the mirror because the architect missed the crack on the original the punch (deficiency) list walk-through.

49. Nine months after final payment, the owner discovers the operation and maintenance manual for a 5-ton air conditioning unit was never received. What should the architect do?
    a. Request the contractor to provide the operation and maintenance manual.
    b. Withhold 5% of the retainage (holdback) until the contractor provides the operation and maintenance manual.
    c. Contact the air conditioning unit manufacturer directly to obtain the operation and maintenance manual.

d. Have the owner contact the air conditioning unit manufacturer directly to obtain the operation and maintenance manual.

50. Which of the following is proper language for an agreement between owner and architect? **Check the two that apply.**
    a. The architect shall perform its service with professional skill and best care available in the same locality.
    b. The architect shall perform its service with professional skill and ordinary care available in the same locality.
    c. The architect shall perform its service expeditiously.
    d. The architect shall perform its service per the term of "time is of essence."

51. A statement of the company's financial activities while excluding "unusual and nonrecurring transactions" when stating how much money the company actually made is called:
    a. financial planning
    b. *pro forma* accounting
    c. revenue accounting
    d. profit/loss accounting

52. MASTERSPEC is a product of:
    a. AIA
    b. CSI
    c. ANSI
    d. ASTM

53. The U.S. National CAD Standard (NCS) promotes "Uniform Drawing Standard" (UDS). UDS uses a coordinate-based location system (grid) to organize information on each drawing sheet. Which of the following is true? **Check the two that apply.**
    a. The drawing grids are numbered from left to right in the order of 1, 2, 3...
    b. The drawing grids are numbered from right to left in the order of 1, 2, 3...
    c. The drawing grids are labeled from top to bottom in the order of A, B, C...
    d. The drawing grids are labeled from bottom to top in the order of A, B, C...

54. Per *ADAAG Manual: A Guide to the American with Disabilities Accessibility Guidelines* by Access Board, the minimum clear passage width for a single wheelchair at a doorway is ________________.

55. Which of the following should occur before painting the interior? **Check the three that apply.**
    a. All interior gypsum boards are installed.
    b. Exterior windows are sealed.
    c. Exterior trees and lawn are planted.
    d. Roofing is completed.
    e. Irrigation lines are installed.
    f. Sidewalks are completed.

56. Which of the following will affect the estimated construction cost of a project? **Check the three that apply.**
    a. The experience of the contractor
    b. The experience of the architect
    c. The number of submittals
    d. The percentage of retainage (holdback)
    e. The number of RFIs
    f. The bidding climate

57. A developer wants to use local contractors to build a shopping center to help stimulate the local economic. The budget is $50 million. None of the local contractors has the bonding capacity of $50 million. Which of the following is a possible solution?
    a. Design-bid-build
    b. Fast-track
    c. Using change orders
    d. Using multiple prime contractors

58. An architect has to prepare several individual bid document packages for which of the following?
    a. Design-bid-build
    b. Design-build
    c. Bridging
    d. Fast-track

59. If an architect signs a standard B101, Standard Form of Agreement between Owner and Architect, which of the following insurance shall she maintain during the duration of the agreement? **Check the four that apply.**
    a. General Liability
    b. Professional Liability
    c. Automobile Liability
    d. Property Liability
    e. Workers' Compensation
    f. Fire Liability

60. Per B101–2007, Standard Form of Agreement between Owner and Architect, which of the following is part of the architect's basic service? **Check the four that apply.**
    a. Electrical engineering
    b. Architectural design
    c. Coordination with the architect's consultants
    d. Project meetings
    e. Construction schedule
    f. Construction sequence

61. What is the difference between life cycle costing (LCC) and life cycle assessment (LCA)?
    a. LCA focuses on economic analysis while LCC focuses on environmental analysis.
    b. LCC focuses on economic analysis while LCA focuses on environmental analysis.
    c. LCA focuses on environmental and economic analysis.
    d. LCC focuses on environmental and economic analysis.

62. A structural engineer specifies pre-fabricated roof trusses for a project. Which of the following is correct? **Check the two that apply.**
    a. The structural engineer should show the roof trusses on the framing plans.
    b. The truss engineer should show the roof trusses on the framing plans.
    c. The structural engineer should prepare, stamp, and sign the trusses plans.
    d. The truss engineer should prepare, stamp, and sign the trusses plans.

63. During construction, a contractor submitted the roof tile substitution and followed up with a phone call the next day to the architect, requesting the architect to approve the roof tile within 2 days. The contractor also informed the architect that if the roof tile substitution was not approved within 2 days, the project would be delayed. What is a proper action for the architect?
    a. Approve the roof tile substitution within 2 days to avoid potential delay of the project.
    b. Reject the roof tile substitution.
    c. Discuss with the owner and let the owner approve or reject the substitution within a reasonable timeframe.
    d. Approve or reject the substitution within a reasonable timeframe.

64. Per B101, Standard Form of Agreement between Owner and Architect, which of the following are additional architectural services? **Check the four that apply.**
    a. A meeting with the owner to review facility operations and performance 9 months after final completion
    b. Preparing Change Orders and Construction Change Directives and supporting documents
    c. Field Inspections to determine date of substantial completion and date of final completion
    d. Programming
    e. Building information modeling
    f. Value Analysis

65. A major fast food chain owner offers an architect an agreement between owner and architect generated by the owner's corporate attorney. The agreement designates the architect's work as "work for hire." What is the implication if the architect signs this agreement?
    a. It simply means the owner has hired the architect as a consultant for the project.
    b. It means the owner has hired the architect and obtained the license to use the architect's plans generated for the project.
    c. It means the owner has hired the architect and obtained the non-exclusive license to use the architect's plans generated for the project.

d. It means the owner has hired the architect and obtained all the rights for the architect's plans generated for the project.

66. If a dispute arises between the owner and the architect, which of the following is a proper procedure to resolve the dispute?
    a. The owner and the architect shall seek arbitration and then mediation and then litigation.
    b. The owner and the architect shall seek mediation and then arbitration and then litigation.
    c. The owner and the architect shall seek mediation and then arbitration.
    d. The owner and the architect shall seek mediation and then arbitration or litigation.

67. Which of the following regarding mediation, arbitration, and litigation is not true?
    a. A mediator is usually a building owner, architect, contractor, or attorney practicing in the construction industry and has judge-like power.
    b. An arbiter is usually a building owner, architect, contractor, or attorney practicing in the construction industry and has judge-like power.
    c. Agreements reached in mediation and arbitration are enforceable in any court having jurisdiction thereof.
    d. Both arbitration and litigation involve a third party with judge-like power.

68. An architect can include by ______________ the mechanical engineer in an arbitration with the owner regarding HVAC issues if the mechanical engineer agrees to the request in writing.

69. An owner is working with a contractor on an owner-contractor agreement, A101, Standard Form of Agreement Between Owner and Contractor, where the basis of payment is a Stipulated Sum. Which of the following is incorrect regarding revising the standard AIA form? **Check the four that apply.**
    a. Use correction fluid to wipe out the languages not used.
    b. Use Xs to completely block out the languages not used.
    c. Use single line to strike out the languages not used.
    d. Retype the AIA document in MS Word format so that it looks clean and reads better.
    e. Both parties shall initial corrections.
    f. Both parties shall initial and seal corrections.

70. A contractor received payment from the owner on March 31, 2011, up to what date did the payment cover work completed?
    a. Around March 1, 2011
    b. Around March 16, 2011
    c. Around March 24, 2011
    d. Around February 1, 2011

71. Which of the following is normally not part of the contract documents? **Check the three that apply.**
    a. General Condition
    b. Shops drawings
    c. Submittals
    d. Addenda
    e. Modifications
    f. Geotechnical reports

72. Which of the following is true regarding the final payment?
    a. The final payment is due to the contractor at substantial completion.
    b. The owner should pay the final payment 30 days after the receipt of the architect's final Certificate of Payment.
    c. If there is a surety for the project, the final payment needs the approval of the surety.
    d. The surety should issue the final Certificate of Payment.
    e. Conditional lien waivers are due before the final payment.
    f. Unconditional lien waivers for the remaining balance are due before the final payment.

73. A building official can enforce all of the following except
    a. the International Building Code
    b. the American with Disabilities Act (ADA)
    c. the Life Safety Code (NFPA 101)
    d. the National Electrical Code (NFPA 70)

74. In which CSI MasterFormat 2004 division can an architect find the specification for stone?
    a. Division 03
    b. Division 04
    c. Division 05
    d. Division 06

75. Per International Code Council, Inc. (ICC), *International Building Code* (IBC), the minimum number of plumbing fixture can be determined by: **(Check the four that apply.)**
    a. number of occupants
    b. number of rooms
    c. number of units
    d. occupancy
    e. tenant's prototype plans
    f. water pressure

76. An architect is doing initial project research. Select the three most important issues that she needs to consider from the following:
    a. zoning
    b. occupancy
    c. exit width
    d. if the building will be sprinklered
    e. plumbing fixtures

f. HVAC equipment

77. A contractor is doing demolition work at a project. She suspects the HVAC ducts inside a concealed attic space contain asbestos. She stops the work and informs the owner. The owner hires a testing lab. The lab confirms the presence of asbestos. The project is stopped for two weeks while the owner hires a specialist to do the asbestos abatement. When the area is clear, the contractor submits a change order request for $6,000 and a time extension of two weeks. The owner rejects the change order. What is the next step?
    a. The dispute should be referred to the architect.
    b. The dispute should be referred to IDM.
    c. The dispute should proceed directly to mediation.
    d. The architect should approve the time extension, and refer the dispute of the cost to mediation.

78. An architect is working with a developer on a mixed-use complex. Which of the following is not an additional service?
    a. As-constructed record drawings
    b. Commissioning
    c. Presentations in a public hearing to defend owner's choice of water fountains and sculpture
    d. Coordination of mechanical and electrical ceiling plans

79. A form of strict, secondary liability that arises under the common law doctrine of agency, the responsibility of the superior for the acts of their subordinate is known as
    a. Employee liability
    b. Vicarious liability
    c. General liability
    d. Limited liability

80. Which of the following items is likely to be shop fabricated?
    a. Plumbing fixtures
    b. HVAC equipment
    c. Storefront
    d. Slab

81. Which of the following statements regarding construction schedule is incorrect?
    a. The contractor is responsible for the construction schedule.
    b. The architect sets up the criteria for the construction schedule.
    c. CPM schedule shows the duration, sequences, and relationship between activities.
    d. The shortest path is the critical path in CPM.

82. An architect contracts with an electrical engineer to provide electrical engineering service for an office building. Who is responsible to the owner for the electrical system's design?
    a. The architect
    b. The electrical engineer
    c. The architect and the electrical engineer jointly

d. The contractor

83. When selecting switchgear, an architect and his electrical engineer should consider which of the following factors? **Check the four most important factors that apply.**
    a. Dimension
    b. Voltage
    c. Noise characteristics of the equipment
    d. Clearance
    e. Available space
    f. Color

84. Per C401, Standard Form of Agreement Between Architect and Consultant, which of the following statements is true?
    a. The consultant shall maintain professional liability insurance for the duration of the agreement.
    b. The consultant shall maintain workers' compensation and professional liability insurance for the duration of the agreement.
    c. The consultant shall maintain automobile liability, workers' compensation, and professional liability insurance for the duration of the agreement.
    d. The consultant shall maintain property liability, automobile liability, workers' compensation, and professional liability insurance for the duration of the agreement.
    e. The consultant shall maintain general liability, automobile liability, workers' compensation, and professional liability insurance for the duration of the agreement.
    f. There are no requirements regarding consultant's insurance

B. Case Study

Questions 85 through 95 refer to the following case study. See figure 3.1 though figure 3.3 for information necessary to answer the questions.

Directions

Fill in the blanks or drag and drop symbols to complete a building section according to the plans and program requirements. The section cut line indicates the location of section. Show grade line, exterior walls, bearing walls, foundations, footings, slabs on grade, finished ceilings, interior walls, ducts, joists, decks, parapets, and roofs. The section must show joists in elevation immediately adjacent to the cut line. The locations of section elements must match the floor plans. The section must accurately show the minimum room heights, parapet heights, and clearance spaces needed for light fixtures and ducts. Use the minimum dimensions and clearance spaces to save on building costs.

Program

The structure system for this building is open web steel joists on masonry bearing walls with continuous concrete spread footing.

1. The slab on grade is 5" thick.
2. All roofs and ceilings are flat.
3. All open web steel joists have 4" thick concrete deck on top.
4. The ceiling height for the multi-purpose room is 18'-0". The ceiling height for the remaining rooms on the 1st floor is 8'-6". The ceiling height for all the rooms on the 2nd floor is 9'-0". The ceiling thickness can be ignored.
5. Corridor walls are 1-hour fire-rated. Transfer grilles and fire/smoke dampers are provided as needed.
6. Exterior walls and interior bearing walls are 2-hour fire-rated.
7. The site is flat.
8. The frost depth is 4'-6". Minimum spread footing size is 12" high x 24" wide.
9. The top of parapet must be 3'-0" above the highest adjacent roof.
10. The clearance space for light fixtures is 8". All ducts must be placed below joists, and all light fixtures must be placed below the ducts and above the ceilings.
11. All ceilings are non-rated, and the ceiling spaces are used as return air plenums.

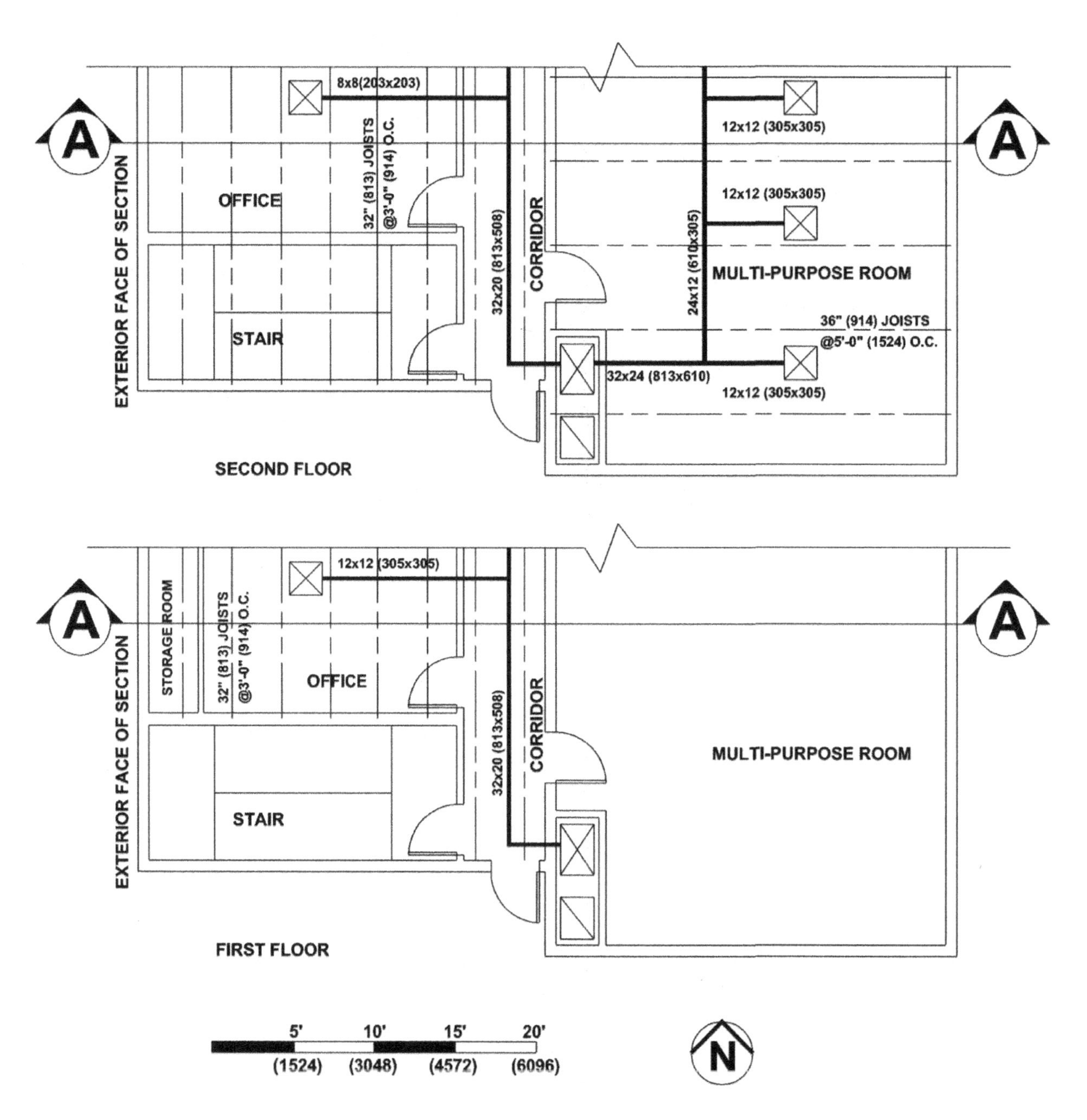

Figure 3.2 Plans for case study

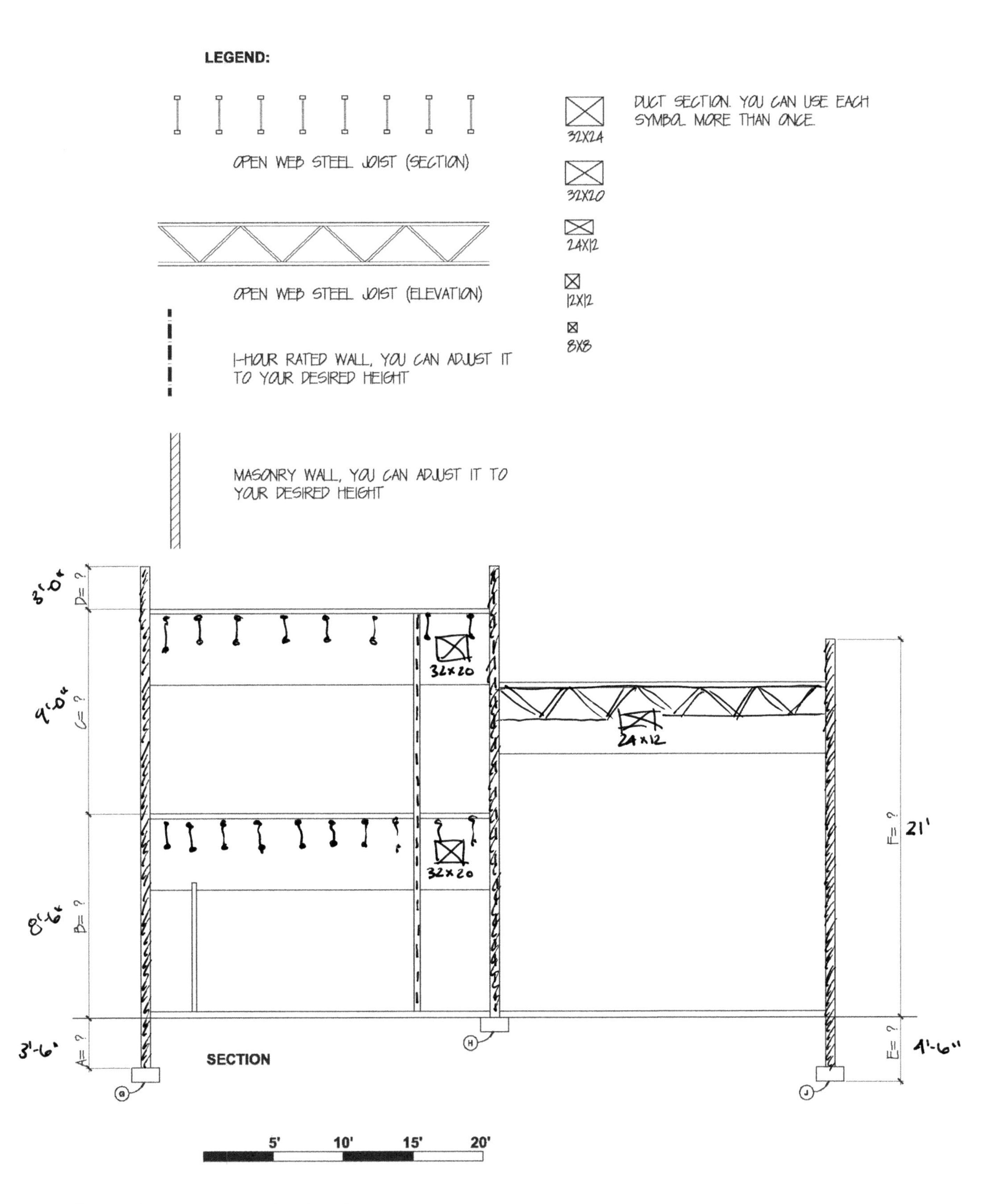

Figure 3.3 Section for case study

85. Drag and place the proper symbols on the section to show the masonry walls and 1-hour fire-rated wall.

86. Drag and place the proper duct symbols on the section.

87. Drag and place the proper symbols on the section to show the open web steel joists per the program.

88. Fill in the blanks to show the dimensions at A, B, C, and D.

89. Fill in the blanks to show the dimensions at E and F.

90. Which of the following regarding footing G, H, and J is correct?
    a. One of the footing depths is incorrect.
    b. The top of the footings must align with or below the frost line.
    c. The bottom of the footings must align with or below the frost line.
    d. none of the above

91. During a site visit, a structural engineer suspected some of the rebar for the footing may not match the plans. She reported this to the architect at the end of the day. The architect called the contractor, and the contractor informed the architect that the concrete footing had been poured. What should the architect do?
    a. Immediately call and inform the owner.
    b. Accept the work but ask the contractor for credit.
    c. Ask the contractor to open the concrete footing to check if the rebar is installed correctly.
    d. Do nothing.

92. An electrical engineer noticed that the light fixtures were not installed are the per the plans, and informed the architect. What should the architect do?
    a. Immediately call and inform the owner.
    b. Ask the contractor to provide a reason.
    c. Accept the work but ask the contractor for credit.
    d. Reject the work.

93. Two weeks before the bid date, the owner decides to use wood trusses instead of open web steel joists. What is the best way to communicate this change to bidders?
    a. an addendum
    b. a notice
    c. a construction change directive (CCD)
    d. a change order

94. The contractor submitted a payment request for the completed concrete floor and footing, as well as the masonry materials stocked off-site. The contractor produced a receipt for the masonry materials. What should the architect do?
    a. Approve the payment request.
    b. Approve the portion for the completed concrete floor and footing, but deny the portion for the masonry materials stocked off-site.
    c. Inform the owner.
    d. Reject the payment request.

95. A subcontracter drops off a box of ceiling tile submittals, and informs the architect that they should be reviewed and approved within two days to avoid project delay. The architect notices that the general contractor has not reviewed and stamped the submittals. What should the architect do?
    a. Review the submittals ASAP to avoid project delay.
    b. Return the submittals to the general contractor without any action.
    c. Inform the owner.
    d. Reject the submittals.

Chapter Four

ARE Mock Exam Solutions for Construction & Evaluation (CE) Division

A. Mock Exam Answers and Explanations: Multiple-Choice (MC)

Note: If you answer 62% of the questions correctly, you will pass the exam.

1. Answer: f
   See IBC, Section 1004 and Table 1004.1.2.
   The occupant load factors are calculated based on the number of fixed seating, the net floor area, or the gross floor area, but subject to the official's final decision. For example, per Table 1004.1.2, the occupant load factor for mercantile use is based on the gross floor area, but the occupant load factor for day care use is based on the net floor area. See Section 1004.1.2 for examples of exceptions that a building official can grant.

   See the following link for *free* IBC code:
   https://codes.iccsafe.org/public/document/code/542/9677240

2. Answer: b
   It is the contractor's responsibility to coordinate all his subcontractors' work and make sure the construction work is performed per the contract documents. The owner has nothing to do with this. The architect's routine job site visits are for field observation and to verify the construction work is in general compliance with the contract documents, but it is still the contractor's responsibility to coordinate all his subcontractors' work and make sure the construction work is performed per the contract documents.

3. Answer: a
   The contractor will pay for the missed item since a general note on the cover sheet covers this item. Per A201, General Conditions of the Contract for Construction, the contract documents are complementary, and what is required by one shall be as binding as if required by all. The contractor should have coordinated with all his subcontractors and submitted the bid based on the entire set of contract documents, not just one portion.

   On the other hand, as an architect, you should also coordinate with your consultants to avoid conflicts within the contract documents. This shows you the importance of coordinating the plans and specifications.

4. Answer: b
   She should continue to look for the stainless steel urinals that meet the requirements.

She can try to talk with the building officials to gain an exception to the ADA and the local building codes, but this request should NOT be based on economic hardship. Using a stainless steel handicap toilet instead of the stainless steel urinals will NOT really provide equal access. The stainless steel urinals installed in other public restrooms in the same geographic area may NOT meet the requirements.

5. Answer: d
   According to Rule 3.3 of the *Rules of Conduct* published by NCARB, she should report the violation to the building officials.

   The *Rules of Conduct* is available as a FREE PDF file at:
   http://www.ncarb.org/

   Rule 3.3 of the *Rules of Conduct* states:
   "If, in the course of his/her work on a project, an architect becomes aware of a decision taken by his/her employer or client, against the architect's advice, which violates applicable state or municipal building laws and regulations and which will, in the architect's judgment, materially and adversely affect the safety to the public of the finished project, the architect shall
   (i) report the decision to the local building inspector or other public official charged with the enforcement of the applicable state or municipal building laws and regulations,
   (ii) refuse to consent to the decision, and
   (iii) in circumstances where the architect reasonably believes that other such decisions will be taken notwithstanding his/her objection, terminate his/her services with reference to the project unless the architect is able to cause the matter to be resolved by other means.

   In the case of a termination in accordance with Clause (iii), the architect shall have no liability to his/her client or employer on account of such termination."

6. Answer:
   The right to use property of another without possessing it is called **Easement**.

7. Answer: a
   Prior to deleting the toilet stalls, the architect should check the requirements in the current local building codes and plumbing codes. The IPC and IBC are NOT enforceable until they are adapted as part of the local codes. The governing agencies need some time to adapt the IPC or IBC, and the latest IPC or IBC may NOT be have been adapted yet.

8. Answer:
   The public school district can use **condemnation via eminent domain (or expropriation)** to purchase the properties at fair market value without the residents' consent.

9. Answer: a, b, c, and d
   Per B101, Standard Form of Agreement Between Owner and Architect (RAIC Document 6), the owner can reduce the project scope, increase the budget by $2 million, ask the architect to revise the design without paying extra fees, and negotiate with the contractor

to reduce the construction cost. She cannot bill the architect's insurance company $2 million, or back charge the architect $2 million.

10. Answer: a, c, and d
Please pay attention to the word "not." Since we are looking for the statements that are NOT true, the following untrue statements are the correct answers:
   - A substitution is acceptable if it is cheaper than the specified product.
   - A substitution can be verbally approved by the owner and the architect.
   - A substitution is acceptable if it is from a local manufacturer.

   The following true statements are the incorrect answers:
   - A substitution can be acceptable if a specified product is not available.
   - A substitution cannot be submitted as part of the shop drawings without any special notation.
   - A substitution must have equal or better quality and performance as the specified product.

11. Answer: a and b
Terms and condition regarding reimbursable expenses and additional services should be part of the written architectural service contract. The other items such as professional liability insurance, type of the construction, and schematic design fee may be part of some written architectural service contracts, but they may not even be needed for other cases.

12. Answer: a and e
Please pay attention to the word "not." Since we are looking for the measures that cannot divert construction waste from landfill, the following are the correct answers:
   - Using one dumpster at the site
   - Implementing the policy of "reduce, reuse, and recycle" in the architect's office (this will reduce waste, but NOT construction waste)

   The following measures are the incorrect answers:
   - Using three dumpsters at the site: trash, plants, and recycle
   - Requiring all packaging materials at the site to be recycled
   - Setting up a construction waste management plan
   - Seeking LEED certification for the building

13. Answer:
In contract law, the doctrine that a contract can only impose obligations or confer rights to parties to the contract is called **privity**.

   See the FREE PDF file of commentary for AIA document A201–2007, General Conditions of the Contract for Construction.

14. Answer: a, c, and e
The following can help assure quality control of the construction documents, and they are the correct answers:

- Allowing adequate time for preparing the construction documents before bidding (This will definitely help assure quality control of the construction documents)
- Having the contractor review the plans (The contractor is an expert in construction and can really help assure quality control of the construction documents)
- Using checklists (A very efficient way of assure quality control of the construction documents)

The following are the incorrect answers:

- Having the owner review the plans (Owner is not an expert in construction and cannot really help assure quality control of the construction documents)
- Having a review meeting with the building officials (Building officials may help you in interpreting building codes, but they are not expert in construction and cannot really help assure quality control of the construction documents)

Using critical path method (This is part of the fast-track method, and has nothing to do with assuring quality control of the construction documents)

15. Answer: c

    The owner shall provide the information to the contractor within 15 days of receiving the request per A201, General Conditions of the Contract for Construction.

16. Answer: a

    Per A201, General Conditions of the Contract for Construction, after the commencement of the work, the owner does NOT have to provide evidence of owner's financial arrangement for the project unless:

    - The owner fails to make the payments per the contract documents
    - A project's scope change affects the contract sum
    - The contractor indicates in writing a reasonable concern regarding owner's abilities to make payment when due.

17. Answer: c

    Matters regarding the rights and responsibilities of owner, contractor and the architect to the contract regarding a <u>specific</u> project are best handled in <u>supplemental condition</u>.

18. Answer: c

    The contractor should pay for the temporary power during the construction according to A201, General Conditions of the Contract for Construction, Section 3.4.1.

19. Answer: c

    The proper action for the contractor is to stop the work at the affected area and notify the owner and the architect in writing immediately, but in no event later than 21 days after the first observance of the concealed condition. See A201, General Conditions of the Contract for Construction, Section 3.7.4 and 10.3.2.

20. Answer: a

    The best way to handle this in the bidding process is to ask the contractors to include the equipment screen wall as an allowance in the bid. See A201, General Conditions of the

Contract for Construction, Section 3.8. A change order and an order for a minor change in the Work are used during construction, not the bidding process.

21. Answer: d
 The cost-plus-fee contract is best for the owner if the project scope or timing is unknown, and high quality is paramount.

 See *The Architect's Handbook of Professional Practice* (AHPP).

22. Answer: d
 The owner should pay for the field tests not included in the contract documents.

23. Answer: a
 During construction, the first step to resolve the dispute between the owner and the contractor is seeking the Initial Decision Maker's (IDM's) interpretation. The IDM is typically the architect unless the owner and the contractor designated another party to be the IDM in the contract documents.

 See FREE PDF file of commentary for AIA document A201, General Conditions of the Contract for Construction at the following link:
 https://www.aiacontracts.org/resources/64171-a201-2007-general-conditions-commentary

24. Answer: a
 The architect will issue a change order to adjust the contract sum to reflect the $500 reduction from the allowance amount.

 The contractor shall bill the owner $9,500. The installation cost, the contractor's overhead, and profit are ALREADY included in the contract sum.

 See the FREE PDF file of commentary for AIA document A201, General Conditions of the Contract for Construction.

25. Answer: a
 Reject the window flashings work. Rejecting work is a primary means for the architect to request the contractor to correct defects or deficiencies in the contractor's work. It is also the best choice for this situation.

 Owner, not the architect, has the authority to stop work.

 See FREE PDF file of commentary for AIA document A201, General Conditions of the Contract for Construction.

26. Answer: a
 The punch list (the list of items to be corrected by the contractor) should be prepared by the contractor.

27. Answer: c
    The construction sequence of footing, concrete slab, framing, roofing, drywall, painting, and carpet is most likely to be productive.

    The key is roofing must happen before drywall, painting, and carpet so that the roof can protect the indoor work from rain or other weather conditions.

28. Answer: b
    The contractor is responsible to the building owner for the electrical subcontractor's death since the contractor is responsible to the owner for the all his work and his subcontractors' work and the safety procedures at the job site. The building owner has a contract with the contractor, NOT the subcontractors. However, the subcontractors are responsible to the contractor for their portion of the work.

29. Answer: a
    The architect should prepare the change orders.

30. Answer: d
    An architect normally conducts the following inspections:
    - Inspection to determine the date or dates of substantial completion
    - Inspection to determine the date of final completion

    The other answers are just **distracters**.

31. Answer: c, d, and f
    The following regarding change orders and construction change directives are true:
    - When the contractor agrees, and signs a construction change directive, it should be effective immediately and should be recorded as a change order.
    - Change orders have to be signed by the owner, contractor, and the architect.
    - Construction change directives have to be signed by the owner and the architect.

    The following regarding change orders and construction change directives are untrue:
    - Change orders are for major changes, and construction change directives are for minor change. (Both can be for major changes.)
    - Both change orders and construction change directives have to be signed by the owner, contractor, and the architect. (Construction change directives do NOT have to be signed by the contractor.)
    - Construction change directives have to be signed by the owner, contractor, and the architect.

    See the FREE PDF file of commentary for AIA document A201, General Conditions of the Contract for Construction.

32. Answer: a.
   The modifications have to be submitted to the surety for approval. Per AIA Document A312, article 8: "The Surety hereby waives notice of any change, including changes of time, to the Construction Contract or to related subcontracts, purchase orders and other obligations." However, A312, article 8 does NOT prevent modifications from being submitted to the surety for approval. Per the AIA Document **A201** Commentary, Section 7.3.1**:** "Modifications that materially alter the scope of the contract **<u>should be</u>** submitted for approval of the surety **<u>to ensure</u>** that the surety will not be released from its obligations by such changes." So, based on all the facts so far, the correct answer is a.

   A surety is the contractor's insurance company since the contractor is paying for the premium, even though the owner is named as the insured. It is different from the owner's insurance company.

33. Answer: a and c
   Boiler and machinery insurance is the owner's responsibility. If there is no boiler and machinery insurance, and the owner does not notice that the contractor does not intend to purchase the boiler and machinery insurance, then the owner effectively becomes the insurer, and she will need to pay for any related damages.

34. Answer: a and d
   A waiver of subrogation should be requested before any loss occurs. The owner has to disclose the subrogation provision to the insurer before purchasing the property insurance.

35. When the slope of a sidewalk is more than **<u>5</u>%** in the direction of travel, the sidewalk becomes a ramp.

36. When you are working on the construction drawings for a wood stud building, you should dimension to the **<u>face</u>** of stud.

37. The five classical orders are **<u>Tuscan, Doric, Ionic, Corinthian, and Composite</u>.**

   You should have learned about them in architectural history classes. You can also find this information in *Architectural Graphic Standards*.

38. Answer: d
   Answer "a" is the symbol for single receptacle outlet, answer "b" is the symbol for floor single receptacle outlet, answer "c" is the symbol for fan hanger receptacle, and answer "d" is the correct answer: the symbol for floor special purpose outlet.

   See *Architectural Graphic Standards*.

39. Answer:
   A developer is building a shopping center. All buildings in this shopping center are one-story. The occupancy group for these building is Group <u>M</u>. The most cost-effective way to design and build these building is to qualify all of them as one "building," Type <u>V</u>

construction, fully sprinklered, with 60-feet-wide public ways or yards around all sides of the shopping center so that the area of all these buildings shall not be limited.

See Section 507.4 and Table 1004.1.1 of *International Building Code* (IBC), by International Code Council, Inc. (ICC).

See the following link for some FREE IBC code sections citations:
https://codes.iccsafe.org/public/document/code/542/9668166

40. When the occupant load for a room is 780, the minimum number of exit(s) required for this room is 3.

According to Table 1006.3.1 of *International Building Code* (IBC), by International Code Council, Inc. (ICC):

Minimum Number of Exits or Access to Exits per Story:

Occupant Load Per Story	Minimum Number of Exits or Access to Exits from Story
1-500	2
501-1,000	3
>1,000	4

41. Answer: b, d, e, and f
The following are project delivery methods:
- Design-Bid-Build
- Construction Management
- Design-Build
- Integrated Project Delivery

BIM is an enabling technology, NOT a project delivery method.

Construction Administration is one stage in construction, NOT a project delivery method. Both BIM and Construction Administration are distracters to confuse you.

See *The Architect's Handbook of Professional Practice* (AHPP).

42. Answer: d
Program management provided by construction management entities works best for complicated projects like an airport. Therefore, the correct answer is construction management.

See *The Architect's Handbook of Professional Practice* (AHPP).

43. Answer: c
If quality is the most important variable for a project, negotiated select team is the preferred project delivery method.

Cost plus fixed fee is the preferred project delivery method if the project scope is not defined clearly.

Design-Build is the preferred project delivery method if the project risk is the most important variable for a project. Bridging is a sub-category of design-build.

See *The Architect's Handbook of Professional Practice* (AHPP).

44. Answer: a
    In a concrete **slump test**, if a shear slump or collapse slump is achieved, the concrete is too wet. The test is NOT acceptable, and a fresh sample should be taken and the test repeated.

    See the following link for more information:
    http://en.wikipedia.org/wiki/Concrete_slump_test

45. Answer: c
    If the owner hires a design architect and a production architect, she is likely to use bridging or bridge design-build.

    Design-Bid-Build is a traditional project delivery method, and often involves one architect.

    Design-Build involves an entity that does both design and construction, and often involves one architect.

    If quality is the most important variable for a project, negotiated select team is the preferred project delivery method. Negotiated select team often involves one architect.

    BIM is an enabling technology, NOT a project delivery method.

    Cost plus fixed fee is the preferred project delivery method if the project scope is not defined clearly. Cost plus fixed fee often involves one architect.

    Design-build includes standard Design-build and bridging. It is a possible answer, but it is NOT as good as bridging or bridge design-build.

    See *The Architect's Handbook of Professional Practice* (AHPP).

46. Answer: d
    Standard design-build is likely to allow the owner to know the project cost at earliest stage of the project: before the Predesign (PD) stage.

    Design-Bid-Build is likely to allow the owner to know the project cost after the Contract Documents (CD) are completed.

    Cost plus fixed fee and negotiated select team are likely to allow the owner to know the project cost after the Schematic Design (SD) stage.

BIM is an enabling technology, NOT a project delivery method.

CM-agent is likely to allow the owner to know the project cost after the Contract Documents (CD) are completed.

See *The Architect's Handbook of Professional Practice* (AHPP).

47. Answer:
The actual dimensions of a 2x4 wood stud are 1 ½" x 3 ½" (38x89).

The actual dimensions of **wood** studs are **smaller** than the nominal dimensions, but the actual dimensions of **metal** studs **match** the nominal dimensions. This difference is important for coordination. For example, the actual dimension of a 4" metal stud is 4" (102). Your structural engineer can place a 4x4 (102x102) tube steel column inside a 4" (102) metal stud wall, but she can NOT place it inside a 2x4 wood stud wall since its actual dimension is 1 ½" x 3 ½" (38x89). When you review your structural engineer's foundation plans and framing plans, you need to pay attention to issues like this.

See the following link for more information:
http://en.wikipedia.org/wiki/Lumber

48. Answer: a
The owner should be responsible for the crack since he has moved in, and the cause of the crack cannot be established.

49. Answer: a
The architect should request the contractor to provide the operation and maintenance manual since this is the contractor's responsibility.

Contacting the air conditioning unit manufacturer directly to obtain the operation and maintenance manual, or having the owner contact the air conditioning unit manufacturer directly to obtain the operation and maintenance manual are possible solutions, but they are not as good as the solution above.

This occurs nine months after final payment. There is no money left in the retainage (holdback).

50. Answer: b and c
The following are proper language for an agreement between owner and architect:
- The architect shall perform its service with professional skill and **ordinary** care available in the same locality.
- The architect shall perform its service expeditiously.

The following are improper language for an agreement between owner and architect:

- The architect shall perform its service with professional skill and **best** care available in the same locality. (This statement will put the architect in a disadvantage position and bring on unnecessary liabilities because it takes a lot more than average time and effort to achieve "best" result. "Best" is also a very subjective term that can be interpreted differently by different people.)
- The architect shall perform its service per the term of "time is of essence." (This statement will put the architect in a disadvantage position and bring on unnecessary liabilities and requires the architect to meet absolute time limitation. There are factors that cannot be controlled by the architect, such as the building department plan check time, etc. A written schedule may be helpful, but it must be subject to changes that are beyond the control of the owner and the architect.)

  See the FREE PDF file of the commentary for AIA documents B101 at the following link for more information:
  https://www.aiacontracts.org/contract-documents/21392-owner-architect-agreement

51. Answer: b
A statement of the company's financial activities while excluding "unusual and nonrecurring transactions" when stating how much money the company actually made is called ***pro forma* accounting**. All other answers are distracters.

See *The Architect's Handbook of Professional Practice* (AHPP), Section 8.4, or the following link for more information:
http://en.wikipedia.org/wiki/Pro_forma

52. Answer: a
MASTERSPEC is a product of AIA, published and supported by ARCOM.

MasterFormat is a product of CSI.

MASTERSPEC uses assigned numbers per MasterFormat. There are related but different.

ANSI and ASTM are distracters.

See *The Architect's Handbook of Professional Practice* (AHPP), Section 12.3.

53. Answer: a and d
The drawing grids are numbered from left to right in the order of 1, 2, 3...

AND
The drawing grids are labeled from bottom to top in the order of A, B, C...

See *The Architect's Handbook of Professional Practice* (AHPP), Section 12.3, or the following links for more information:
http://www.NationalCADStandard.org

OR
http://www.buildingsmartalliance.org/index.php/ncs/content/

54. Answer:
Per *ADAAG Manual: A Guide to the American with Disabilities Accessibility Guidelines* by Access Board, the minimum clear passage width for a single wheelchair at a doorway is 32".

You need to become familiar with all the Accessibility diagrams and critical dimensions at the following link:
https://www.access-board.gov/guidelines-and-standards/buildings-and-sites/about-the-ada-standards

55. Answer: a, b, and d
The following should occur before painting interior (The explanations are in parentheses below. Typical for the rest of the answers):
- All interior gypsum boards are installed.
- Exterior windows are sealed (This will prevent water or moisture damage if it rains).
- Roofing is completed (This will prevent water damage if it rains).

The following can occur after painting interior:
- Exterior trees and lawn are planted.
- Irrigation lines are installed.
- Sidewalks are completed.

56. Answer: a, b, and f
The following will affect the estimated construction cost of a project:
- The experience of the contractor
- The experience of the architect (An architect with good experience can select better building systems, produce high quality plans and specifications, and reduce the construction cost of a project.)
- The bidding climate (It means the anticipated weather conditions during the duration of a project.)

The following will not affect the estimated construction cost of a project:
- The number of submittals (The contractor shall construct the project per the contract documents, even if some of the items do not require submittals.)
- The percentage of retainage (holdback)
- The number of RFIs (Some contractors tend to send in many Requests for Information, or RFIs, and hope to turn some of the RFIs into change orders. If an item is already covered in the contract documents, then it will NOT be a change order item even if it is on the RFI. It is important for an architect to tie the RFI responses back to the contract documents to reduce potential change orders. The contractor shall construct the project per the contract documents.)

57. Answer: d
    Using multiple prime contractors is a possible solution. The other three choices will NOT solve the problem.

58. Answer: d
    An architect has to prepare several individual bid document packages for fast-track projects, such as a civil bid document package, foundation and superstructure bid document package, building enclosure bid document package, interiors bid document package, etc.

    The other three delivery methods (design-bid-build, design-build, including bridging) have one construction contract.

    See *The Architect's Handbook of Professional Practice* (AHPP) for more information.

59. Answer: a, b, c, and e
    If an architect signs a standard B101–2007 (Former B141–1997), Standard Form of Agreement between Owner and Architect (RAIC Document 6), she shall maintain the following insurance during the duration of the agreement:
    - General Liability
    - Professional Liability
    - Automobile Liability
    - Workers' Compensation

    See the FREE PDF file of the commentary for B101–2007 (Former B141–1997), Standard Form of Agreement between Owner and Architect (RAIC Document 6) at the following link:
    http://www.aia.org/contractdocs/aiab081438

60. Answer: a, b, c, and d
    Per B101–2007 (Former B141–1997), Standard Form of Agreement between Owner and Architect (RAIC Document 6), an architect's **basic** services include the following:
    - Electrical, Mechanical, and Structural engineering (An architect can hire engineers as sub-consultants to perform the work.)
    - Architectural design
    - Coordination with the architect's consultants
    - Project meetings
    - Complying with the design requirements of the utilities companies and governing agencies
    - Feasibility of incorporating environmentally responsible design

    An architect's basic services do NOT include the following:
    - Construction schedule (Part of the contractor's work)
    - Construction sequence (Part of the contractor's work)

See Article 3, Scope of the Architect's Basic Services in the FREE PDF file of the commentary for B101, Standard Form of Agreement between Owner and Architect at the following link:
https://www.aiacontracts.org/contract-documents/21392-owner-architect-agreement

61. Answer: b
LCC focuses on economic analysis while LCA focuses on environmental analysis.

See *The Architect's Handbook of Professional Practice* (AHPP) for LCC.

62. Answer: a and d
The following is correct:
- The structural engineer should show the roof trusses on the framing plans.
- The truss engineer should prepare, stamp, and sign the trusses plans. (Trusses are typically design-build by the truss engineer retained by the contractor.)

See the FREE PDF file of the commentary for B101, Standard Form of Agreement between Owner and Architect at the following link:
http://www.aia.org/contractdocs/aiab081438

63. Answer: c
Discuss with the owner and have the owner approve or reject the substitution within a reasonable timeframe. The owner, NOT the architect, has the right to approve or reject the substitution. If the owner approves the substitution, the architect can issue an order for minor changes in the Work, a Change Order, or Construction Change Directive regarding this substitution.

In most cases, the contractor requests substitutions because they are cheaper; an architect should review the request and request proper credit from the contractor if applicable.

The roof tiles are a long-lead item, and the contractor should have ordered them well in advance. Any delay because of the substitution request should be the responsibility of the contractor.

See the FREE PDF file of commentary for AIA document A201, General Conditions of the Contract for Construction, for related information.

64. Answer: b, d, e, and f
Per B101, Standard Form of Agreement between Owner and Architect, the following are **additional** architectural services:
- Preparing Change Orders and Construction Change Directives and supporting documents (Supporting documents are provided by the contractor, the architect can provide them as an additional architectural service)
- Programming
- Building information modeling
- Value Analysis

The following are **basic** architectural services:

- A meeting with the owner to review facility operations and performance 9 months after final completion
- Field Inspections to determine date of substantial completion and date of final completion

See Article 3 and Article 4 of the FREE PDF file of the commentary for B101, Standard Form of Agreement between Owner and Architect at the following link:
http://www.aia.org/contractdocs/aiab081438

65. Answer: d
In the US, the term "work for hire" means the owner has hired the architect and obtained ALL the rights for the architect's plans and other documents generated for the project.

Once an architect transfers all the rights for plans and other documents generated for a project, she cannot even re-use them for another project herself. Make sure you know the implication when you review any contract that deviates from the AIA documents.

The standard B101, Standard Form of Agreement between Owner and Architect (RAIC Document 6) clearly defines that all plans and other documents are instrument of service and the architect owns all the rights and grants a non-exclusive license right (NOT copyright) for the owner to use them for this project ONLY. If the owner wants to use the plans for another project, he needs pay the architect an additional license fee and release and indemnify the architect for such use.

See Article 7 of the FREE PDF file of the commentary for B101, Standard Form of Agreement between Owner and Architect.

66. Answer: d
The owner and the architect shall seek mediation, and then arbitration or litigation.

See Article 8 of the FREE PDF file of the commentary for B101, Standard Form of Agreement between Owner and Architect.

67. Answer: a
A mediator is usually a building owner, architect, contractor, or attorney practicing in the construction industry and has judge-like power. (A mediator does **NOT** have judge-like power and can**not** impose settlement terms.)

Please note we are looking for the **incorrect** statement as the correct answer.

The following are **correct** statements, or the **incorrect** answers:

- An arbiter (or arbitrator) is usually a building owner, architect, contractor. or attorney practicing in the construction industry and has judge-like power.

- Agreements reached in mediation and arbitration are enforceable in any court having jurisdiction thereof.
- Both arbitration and litigation involve a third party with judge-like power.

See Article 8 of the FREE PDF file of the commentary for B101, Standard Form of Agreement between Owner and Architect.

68. Answer:
An architect can include by **joinder** the mechanical engineer in an arbitration with the owner regarding HVAC issues if the mechanical engineer agrees to the request in writing.

See Article 8 of the FREE PDF file of the commentary for B101, Standard Form of Agreement between Owner and Architect.

69. Answer: a, b, d, and f

Please note we are looking for the **incorrect** statement as the correct answer.

The following are incorrect statements, or the correct answers:
- Use correction fluid to wipe out the languages not used. (No, it may raise suspicion of fraudulent concealment.)
- Use Xs to completely block out the languages not used. (No, it may raise suspicion of fraudulent concealment.)
- Retype the AIA document in MS Word format so that it looks clean and reads better. (No, it can introduce typographic errors and cloud legal interpretation of a standard clause, and it is a copyright infringement.)
- Both parties shall initial and seal corrections. (No, no need to seal.)

The following are correct statements, or the incorrect answers:
- Use single line to strike out the languages not used.
- Both parties shall initial corrections.

See Instructions for A101, Standard Form of Agreement Between Owner and Contractor where the basis of payment is a Stipulated Sum.

70. Answer: b
The payment covered work completed up to or around March 16, 2011, because it takes the architect around 7 days to review the application, and the owner around 7 days to pay the contractor.

See Article 5.1.3, Instructions for A101, Standard Form of Agreement Between Owner and Contractor where the basis of payment is a Stipulated Sum.

71. Answer: b, c, and f

Please note we are looking for a document that is normally **not** part of the contract documents as the correct answer.

The following are normally not part of the contract documents, or the correct answers:

- Shops drawings
- Submittals
- Geotechnical reports

The following are normally part of the contract documents, or the incorrect answers:

- General Condition
- Addenda
- Modifications

See A101, Standard Form of Agreement Between Owner and Contractor where the basis of payment is a Stipulated Sum.

72. Answer: c and e.

The following are true regarding the final payment:

- If there is a surety for the project, the final payment needs the approval of the surety.
- Conditional lien waivers are due before the final payment.

The following are not true regarding the final payment:

- The final payment is due to the contractor at substantial completion. (The final payment is due to the contractor at the final completion.)
- The owner should pay the final payment 30 days after the receipt of the Architect's final Certificate of Payment. (The owner should pay the final payment no later than 30 days after the **issuance** of the Architect's final Certificate of Payment.)
- The surety should issue the final Certificate of Payment. (The architect should issue the final Certificate of Payment.)
- Unconditional lien waivers for the remaining balance are due before the final payment. (Unconditional lien waivers for the remaining balance are due after the final payment.)

73. Answer: b

Pay attention to the word "except": A building official can *not* enforce the American with Disabilities Act (ADA) because it is a civil rights registration, *not* a building code. ADA can be a base for civil litigation, and is enforced only by the Department of Justice (DOJ). The other choices are model codes that can be adopted by a local governing agency to become building codes.

A building official can enforce all of the following:

- the International Building Code
- the Life Safety Code (NFPA 101)
- the National Electrical Code (NFPA 70)

74. Answer: b

An architect finds the specification for stone in which Division 04 of MasterFormat 2004 division.

Per MasterFormat 2004, the divisions mentioned in this question are as follows:

- Division 03 Concrete
- Division 04 Masonry (Including 04 50 00 Stone)
- Division 05 Metals
- Division 06 Wood, Plastics, and Composites

You should become familiar with MasterFormat 2004. My other book, *Building Construction*, has detailed discussions on CSI MasterFormat specification sections.

75. Answer: a, b, c, and d

Per International Code Council, Inc. (ICC), *International Building Code* (IBC), the minimum number of plumbing fixture can be determined by:

- number of occupants
- number of rooms (See Institutional Occupancy Group)
- number of units (See Residential Occupancy Group)
- occupancy

Tenant's prototype plans can determine the minimum number of plumbing fixtures for a project, but it is NOT a part of IBC. Water pressure is just a distracter.

See Table 2902.1 of IBC at the following link:
https://codes.iccsafe.org/public/document/code/542/9695828

You should become familiar with Table 2902.1 of IBC. It will help you pass the ARE and help you in your real architectural practice.

76. Answer: a, b, and d

When an architect is doing initial project research, the three most important issues that she needs to consider are:

- zoning
- occupancy
- if the building will be sprinklered

Exit width, plumbing fixtures, and HVAC equipment are less important and can be handled later.

77. Answer: c

The dispute should proceed directly to mediation. Any disputes related to **hazardous** materials should proceed directly to mediation. They are NOT referred to the IDM or the architect.

See comment of Article 10.3.2 of the FREE PDF file of commentary for AIA document A201, General Conditions of the Contract for Construction.

78. Answer: d
Coordination of mechanical and electrical ceiling plans is a basic service for the architect, *not* an additional service.

The following are additional service:
- As-constructed record drawings
- Commissioning
- Presentations in a public hearing to defend the owner's choice of water fountains and sculpture

See Article 4.1 of the FREE PDF file of the commentary for B101, Standard Form of Agreement between Owner and Architect.

79. Answer: b
A form of strict, secondary liability that arises under the common law doctrine of agency, the responsibility of the superior for the acts of his subordinate is known as **vicarious liability**. Employee liability, parental liability, and principal liability are some examples of vicarious liability.

See Introduction section of the FREE PDF file of commentary for AIA document A201, General Conditions of the Contract for Construction.

AND
the definition of vicarious liability at the following link:
http://en.wikipedia.org/wiki/Vicarious_liability

80. Answer: c
Storefront is likely to be shop fabricated and installed at the site. An architect typically reviews shop drawings for storefront.

Plumbing fixtures and HVAC equipment are mass-produced manufactured items.

Slab is placed at the site.

81. Answer: d
Please note we are looking for the incorrect statement as the correct answer.
The following statement regarding construction schedule is incorrect and is therefore the correct answer:
- The **shortest** path is the critical path in CPM.

The following statements regarding construction schedule are correct and are therefore the incorrect answers:
- The contractor is responsible for the construction schedule.

- The architect sets up the criteria for the construction schedule.
- CPM schedule shows the duration, sequences, and relationship between activities.

82. Answer: a

The architect is responsible to the owner for all his work and his consultants' work, including the electrical systems design. The owner has a contract with the architect, NOT his consultants. However, the architect's consultants are responsible to the architect for their portion of the work.

83. Answer: a, b, d, and e

When selecting switchgear, an architect and his electrical engineer should consider the following four most important factors:

- Dimension
- Voltage
- Clearance
- Available space

The following two factors are less important:

- Noise characteristics of the equipment
- Color

84. Answer: e

Per C401, Standard Form of Agreement Between Architect and Consultant, the consultant shall maintain general liability, automobile liability, workers' compensation, and professional liability insurance for the duration of the agreement.

This is one of changes from the previous version of the form, C141, Standard Form of Agreement Between Architect and Consultant.

One way to remember this:

An architect's contract and insurance requirements will "pass down" to her consultants, i.e., consultants need to carry the same kinds of insurance as the architect.

B. Mock Exam Answers and Explanations: Case Study

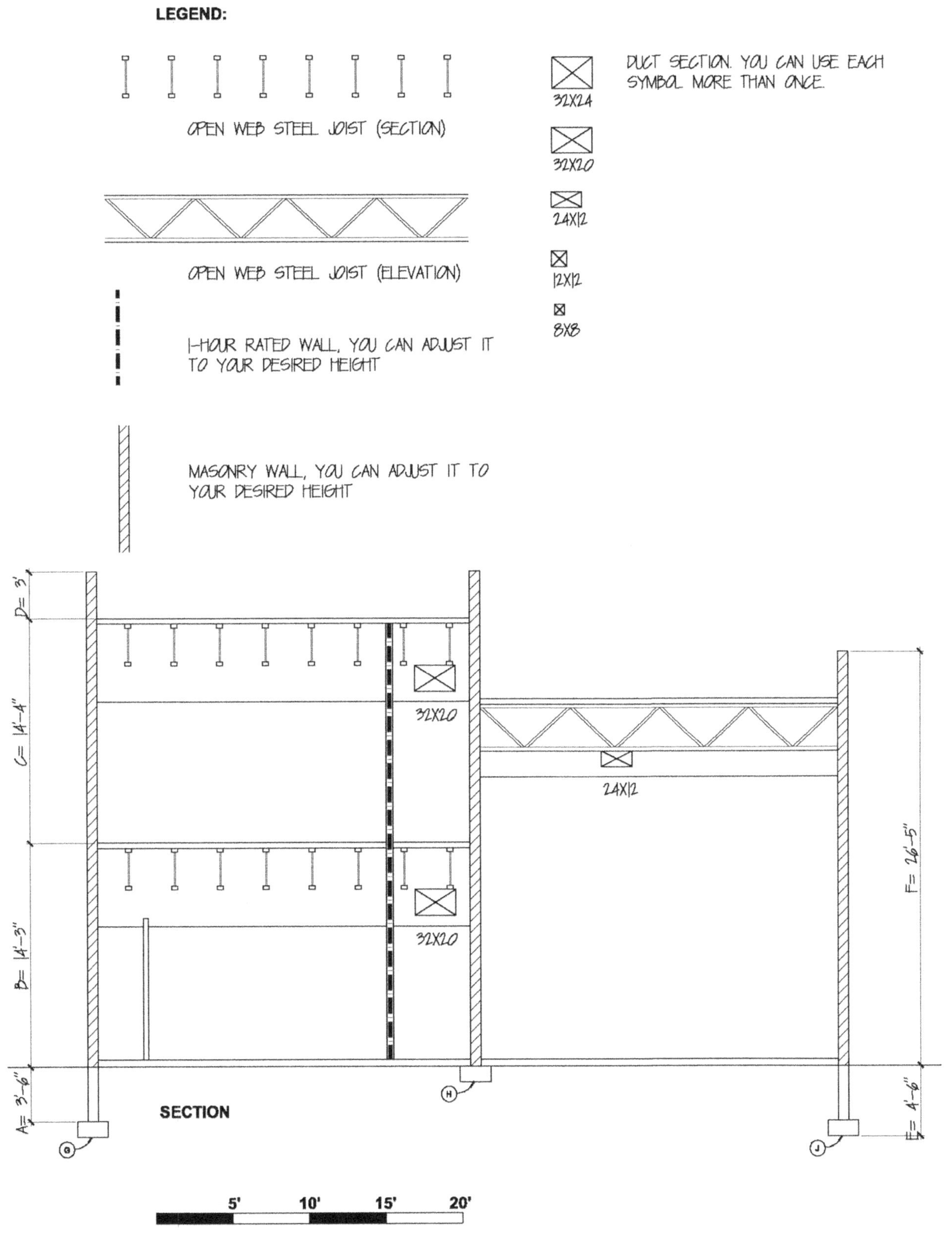

Figure 4.1 Section for case study

85. Answer: See figure 4.1 for the masonry walls and 1-hour fire-rated wall. Per the program, corridor walls are 1-hour fire-rated. You can use the floor plans to project the masonry walls and 1-hour fire-rated wall onto the section.

86. Answer: See figure 4.1 for the ducts on the section. One 32x20 duct is cut above the first floor corridor ceiling, and the other 32x20 duct is cut above the second floor corridor ceiling. One 24x12 duct is cut above the multi-purpose room. You can use the floor plans to project the ducts onto the section.

87. Answer: See figure 4.1 for the open web steel joists. You can use the floor plans to project the open web steel joists onto the section. Make sure you show the joist section symbols at locations where the open web steel joists are cut, align the joist section symbols with the joists on the floor plans, and show a joist elevation symbol for open web steel joists beyond.

88. Answer: See figure 4.1 for the dimensions.

    Note regarding calculation of dimension A:

    1. *The bottom instead of the top of the exterior footings should be placed at frost line. The reasons are as follows:*

       Most soils expand and heave when they freeze.

       *The **frost line** is the depth at which the moisture present in the soil is expected to freeze. Once the bottom of the footing is placed at the frost line the ground will act as a barrier to insulate the soil below the footing from freezing in the winter, and thus prevent the soils below the frost line from expanding and heaving.*

       If you place the top of the footing at the frost line, it will still work, but it will cost MORE money, and it is not as good as placing the bottom of the footing at the frost line. You are supposed to show the minimum depth for the footings.

    2. *The top of the foundation with footings, NOT the top of the slab, should align with the grade line (the bottom of the slab). The reason is:*

       If you align the top of the foundation with footings with the top of the slab, and you draw a 4'-6" deep foundation with footing, your bottom of the footing will be a few inches higher than the frost line and because of the slab thickness, it will not work.

The following are the dimensions:

A = 4'-6" (minimum frost depth) - 12" (minimum spread footing height) = 3'-6"

B = 5" (slab on grade thickness) + 8'-6" (ceiling height) + 8" (clearance space for light fixtures) + 20" (duct height per first floor plan) + 32" (open web steel joist height per first floor plan) + 4" (thickness of the concrete deck on top of the open web steel joists) = 14'-3"

C = 9'-0" (ceiling height) + 8" (clearance space for light fixtures) + 20" (duct height per second floor plan) + 32" (open web steel joist height per second floor plan) + 4" (thickness of the concrete deck on top of the open web steel joists) = 14'-4"

D = 3'-0" (The top of parapet must be 3'-0" above the highest adjacent roof.)

89. Answer: See figure 4.1 for the dimensions.

The following are the dimensions:
E = 4'-6" (minimum frost depth)
F = 5" (slab on grade thickness) + 18'-0" (ceiling height for the multi-purpose room) + 8" (clearance space for light fixtures) + 12" (duct height per second floor plan) + 36" (open web steel joist height per second floor plan) + 4" (thickness of the concrete deck on top of the open web steel joists) + 3'-0" (The top of parapet must be 3'-0" above the highest adjacent roof.) = 26'-5"

90. Answer: c
The bottom of the footings must align with or below the frost line. Once the bottom of the footing is placed at the frost line the ground will act as a barrier to insulate the soil below the footing from freezing in the winter, and thus prevent the soils below the frost line from expanding and heaving.

91. Answer: c
The architect should ask the contractor to open the concrete footing to check if the rebar is installed correctly.

Per A201, General Conditions of the Contract for Construction, if the rebar is installed correctly per the construction documents, then the architect will issue a change order and the owner will pay for the costs of uncovering and replacement. If the rebar is *not* installed correctly, then the contractor will pay for the costs of uncovering and replacement.

92. Answer: d
The architect should reject the work.

An architect can reject the work if it does not comply with the construction documents. This is a major means for the architect to make sure the contractor performs the work according to the construction documents.

93. Answer: a
The best way to communicate this change to bidders is an addendum.

- An **addendum** is used to issue changes that affect time and money before the bid date.
- A **notice** is used to issue changes that do not affect time and money before the bid date.
- A **construction change directive (CCD)** is used to keep project on schedule when the owner and contractor cannot agree on the cost or extra time caused by the changes.
- A **change order** is used to issue changes that affect time and money after the bid date.

94. Answer: b
    The architect should approve the portion for completed concrete floor and footing, but deny the portion for the masonry materials stocked off-site. Materials stocked on-site can be included in payment request, but materials stocked off-site should not be included in payment request.

95. Answer: b
    The architect should return the submittals to the general contractor without any action because the general contractor has not reviewed and stamped the submittals yet.

C. How We Came Up with the Construction & Evaluation (CE) Mock Exam Questions

We came up with all the CE Mock Exam questions based on the ARE 5.0 Handbook, and we developed the Mock Exam based on the *five* weighted sections. See a detailed breakdown in the following tables:

Note: *If the text on following tables is too small for you to read, then you can go to our forum, sign up for a free account, and download the FREE 11x17 full size jpeg format files for these tables at:*
GeeForum.com

Sections	Expected Number of Items	Actual Number of Items
Total	**95**	**95**
Section 1: Preconstruction Activities (17-23%)	**16-22**	**18**
• Interpret the architect's roles and responsibilities during preconstruction, based on delivery method (U/A)		4
• Analyze criteria for selecting contractors (A/E)		5
• Analyze aspects of the contract or design to adjust project costs (A/E)		9
Section 2: Construction Observation (32-38%)	**30-36**	**33**
• Evaluate the architect's role during construction activities (A/E)		15
• Evaluate construction conformance with contract documents, codes, regulations, and sustainability requirements (A/E)		11
• Determine construction progress (U/A)		7
Section 3: Administrative Procedures & Protocols (32-38%)	**30-36**	**34**
• Determine appropriate additional information to supplement contract documents (U/A)		12
• Evaluate submittals including shop drawings, samples, mock-ups, product data, and test results (A/E)		7
• Evaluate the contractor's application for payment (A/E)		7
• Evaluate responses to non-conformance with contract documents (A/E)		8
Section 4: Project Closeout & Evaluation (7-13%)	**6-12**	**10**
• Apply procedural concepts to complete close-out activities (U/A)		6
• Evaluate building design and performance (A/E)		4

Appendixes

A. List of Figures

B. Official reference materials suggested by NCARB

1. Resources Available While Testing

Tips:

- *You need to read through these pages several times and become very familiar with them to save time in the real ARE exams.*

United States. American Institute of Steel Construction, Inc. *Steel Construction Manual*; 14th edition. Chicago, Illinois, 2011.

Beam Diagrams and Formulas:

- Simple Beam: Diagrams and Formulas - Conditions 1-3, page 3-213; Conditions 4-6, page 3-214; Conditions 7-9, page 3-215
- Beam Fixed at Both Ends: Diagrams and Formulas - Conditions 15-17, page 3-218
- Beam Overhanging One Support: Diagrams and Formulas - Conditions 24-28, pages 3-221 & 222

Dimensions and Properties:

- W Shapes 44 thru 27: Dimensions and Properties, pages 1-12 thru 17
- W Shapes 24 thru W14x145: Dimensions and Properties, pages 1-18 thru 23
- W Shapes 14x132 thru W4: Dimensions and Properties, pages 1-24 thru 29
- C Shapes: Dimensions and Properties, pages 1-36 & 37
- Angles: Properties, pages 1-42 thru 49
- Rectangular HSS: Dimensions and Properties, pages 1-74 thru 91
- Square HSS: Dimensions and Properties, pages 1-92 thru 95
- Round HSS: Dimensions and Properties, pages 1-96 thru 100

United States. International Code Council, Inc. *2012 International Building Code*. Country Club Hills, Illinois, 2011.

Live and Concentrated Loads:

- Uniform and Concentrated Loads: IBC Table 1607.1, pages 340-341

2. **Typical Beam Nomenclature**

The following typical beam nomenclature is excerpted from:
United States. American Institute of Steel Construction, Inc. *Steel Construction Manual*; 14th edition. Chicago, Illinois, 2011.

E	Modulus of Elasticity of steel at 29,000 ksi	V_2	Vertical shear at right reaction point, or to left of intermediate reaction of beam, kips
I	Moment of Inertia of beam, in^4	V_3	Vertical shear at right reaction point, or to right of intermediate reaction of beam, kips
L	Total length of beam between reaction point, ft	V_x	Vertical shear at distance x from end of beam, kips
M_{max}	Maximum moment, kip-in	W	Total load on beam, kips
M_1	Maximum moment in left section of beam, kip-in	A	Measured distance along beam, in
M_2	Maximum moment in right section of beam, kip-in	B	Measured distance along beam which may be greater or less than a, in
M_3	Maximum positive moment in beam with combined end moment conditions, kip-in	L	Total length of beam between reaction points, in
M_x	Maximum at distance x from end of beam, kip-in	W	Uniformly distributed load per unit of length, kips/in
P	Concentrated load, kips	w_1	Uniformly distributed load per unit of length nearest left reaction, kips/in
P_1	Concentrated load nearest left reaction, kips	w_2	Uniformly distributed load per unit of length nearest right reaction and of different magnitude than w1, kips/in
P_2	Concentrated load nearest right reaction and of different magnitude than P_1, kips	X	Any distance measured along beam from left reaction, in
R	End beam reaction for any condition of symmetrical loading, Kips	x_1	Any distance measured along overhang section of beam from nearest reaction point, in
R_1	Left end beam reaction, kips	Δ_{max}	Maximum deflection, in
R_2	Right end or intermediate beam reaction, kips	Δa	Deflection at point of load, in
R_3	Right end beam reaction, kips	Δx	Deflection at point x distance from left reaction, in
V	Maximum vertical shear for any condition of symmetrical loading, kips	Δx_1	Deflection of overhang section of beam at any distance from nearest reaction point, in
V_1	Maximum vertical shear in left section of beam, kips		

3. Formulas Available While Testing

Tips:

- *These formulas and references will be available during the real exam. You should read through them a few times before the exam to become familiar with them. This will save you a lot of time during the real exam, and will help you solve structural calculations and other problems.*

Structural:

Flexural stress at extreme fiber

$$f = \frac{Mc}{I} = \frac{M}{S}$$

Flexural stress at any fiber

$$f = \frac{My}{I}$$

where y = distance from neutral axis to fiber

Average vertical shear

$$v = \frac{V}{A} = \frac{V}{dt}$$

for beams and girders

Horizontal shearing stress at any section A-A

$$v = \frac{VQ}{Ib}$$

where Q = statical moment about the neutral axis of the entire section of that portion of the cross-section lying outside of section A-A

b = width at section A-A

Electrical

$$Foot-candles = \frac{lumens}{area\ in\ ft^2}$$

$$Foot-candles = \frac{(lamp\ lumens)\ x\ (lamps\ per\ fixture)\ x\ (number\ of\ fixtures)\ x\ (CU)\ x\ (LLF)}{area\ in\ ft^2}$$

$$Number\ of\ luminaires = \frac{(foot-candles)\ x\ (floor\ area)}{(lumens)\ x\ (CU)\ x\ (LLF)}$$

where CU = coefficient of utilization

LLF = Light Loss Factor

$$DF_{AV} = 0.2x\frac{window\ area}{floor\ area}$$
for spaces with sidelighting or toplighting with vertical monitors

$$\textbf{watts} = \textbf{volts x amperes x power factor}$$
for AC circuits only

$$Demand\ charge = maximum\ power\ demand\ x\ demand\ tariff$$

Plumbing

$$1\ psi = 2.31\ feet\ of\ water$$

$$1\ cubic\ foot = 7.5\ U.S.gallons$$

HVAC

$$\frac{BTU}{year} = peak\ heat\ loss\ x\ \frac{full - load\ hours}{year}$$

$$BTU/h = (cfm)\ x\ (1.08)\ x\ (\Delta T)$$

$$1\ kWh = 3{,}400\ BTU/h$$

$$1\ ton\ of\ air\ conditioning = 12{,}000\ BTU/h$$

$BTU/h = (U)\ x\ (A)\ x\ (T_d)$ *where Td is the difference between indoor and outdoor temperatures*

$$U = 1/R_t$$

$$U_o = \frac{(U_w \times A_w) + (U_{op} \times A_{op})}{Ao}$$
where o = total wall, w = window, and op = opaque wall

$$U_o = \frac{(U_R \times A_R) + (U_S \times A_S)}{Ao}$$

where o = total roof, R = roof, and S = skylight

$$R = x/k$$

where x = thickness of material in inches

$$Heat\ required = \frac{BTU/h}{temperature\ differential} \times (24\ hours) \times (DD\ °F)$$

where DD = degree days

Acoustics

$$\lambda = \frac{c}{f}$$

where λ = wavelength of sound (ft)
c = velocity of sound (fps)
f = frequency of sound (Hz)

$$a = SAC\ x\ S$$

where a = Absorption of a material used in space (sabins)
SAC = Sound Absorption Coefficient of the material
S = Exposed surface area of the material (ft^2)

$$A = \Sigma a$$

Where A = Total sound absorption of a room (sabins)
$\Sigma a = (S_1 \ x \ SAC_1) + (S_2 \ x \ SAC_2) + \ldots$

$$T = 0.05 \times \frac{V}{A}$$

where T = Reverberation time (seconds)
V = Volume of space (ft^3)

$$NRC = average\ SAC\ for\ frequency\ bands\ 250, 500, 1000, and\ 2000\ Hz$$

4. Common Abbreviations

Tips:

- *You need to read through these common abbreviations several times and become very familiar with them to save time in the real ARE exams.*

Professional Organizations, Societies, and Agencies

American Concrete Institute	ACI
American Institute of Architects	AIA
American Institute of Steel Construction	AISC
American National Standards Institute	ANSI
American Society for Testing and Materials	ASTM
American Society of Civil Engineers	ASCE
American Society of Heating, Refrigerating, and Air-Conditioning Engineers	ASHRAE
American Society of Mechanical Engineers	ASME
American Society of Plumbing Engineers	ASPE
Architectural Woodwork Institute	AWI
Construction Specifications Institute	CSI
Department of Housing and Urban Development	HUD
Environmental Protection Agency	EPA
Federal Emergency Management Agency	FEMA
National Fire Protection Association	NFPA
Occupational Safety and Health Administration	OSHA
U.S. Green Building Council	USGBC

Tips:

- *You need to look through the following codes and regulations & AIA contract documents several times and become very familiar with them to save time in the real ARE exams. Read some of the important sections in details.*

AIA Contract Documents

A101-2007, Standard Form of Agreement Between Owner and Contractor - Stipulated Sum	A101
A201-2007, General Conditions of the Contract for Construction	A201
A305-1986, Contractor's Qualification Statement	A305
A701-1997, Instructions to Bidders	A701
B101-2007, Standard Form of Agreement Between Owner and Architect	B101
C401-2007, Standard Form of Agreement Between Architect and Consultant	C401
G701-2001, Change Order	G701
G702-1992, Application and Certificate for Payment	G702
G703-1992, Continuation Sheet	G703
G704-2000, Certificate of Substantial Completion	G704

Codes and Regulations

ADA Standards for Accessible Design	ADA

International Code Council	ICC
International Building Code	IBC
International Energy Conservation Code	IECC
International Existing Building Code	IEBC
International Mechanical Code	IMC
International Plumbing Code	IPC
International Residential Code	IRC
Leadership in Energy and Environmental Design	LEED
National Electrical Code	NEC

Commonly Used Terms

Air Handling Unit	AHU
Authority Having Jurisdiction	AHJ
Building Information Modeling	BIM
Concrete Masonry Unit	CMU
Contract Administration	CA
Construction Document	CD
Dead Load	DL
Design Development	DD
Exterior Insulation and Finish System	EIFS
Furniture, Furnishings & Equipment	FF&E
Floor Area Ratio	FAR
Heating, Ventilating, and Air Conditioning	HVAC
Insulating Glass Unit	IGU
Indoor Air Quality	IAQ
Indoor Environmental Quality	IEQ
Live Load	LL
Material Safety Data Sheets	MSDS
Photovoltaic	PV
Reflected Ceiling Plan	RCP
Schematic Design	SD
Variable Air Volume	VAV
Volatile Organic Compound	VOC
British Thermal Unit	btu
Cubic Feet per Minute	cfm
Cubic Feet per Second	cfs
Cubic Foot	cu. ft. ft^3
Cubic Inch	cu. in. in^3
Cubic Yard	cu. yd. yd^3
Decibel	dB
Foot	ft
Foot-candle	fc
Gross Square Feet	gsf

Impact Insulation Class	IIC
Inch	in
Net Square Feet	nsf
Noise Reduction Coefficient	NRC
Pound	lb
Pounds per Linear Foot	plf
Pounds per Square Foot	psf
Pounds per Square Inch	psi
Sound Transmission Class	STC
Square Foot	sq. ft. sf ft^2
Square Inch	sq. in. in^2
Square Yard	sq. yd.

5. General NCARB reference materials for ARE:

Per NCARB, all candidates should become familiar with the latest version of the following codes:

International Code Council, Inc. (ICC)
International Building Code
International Mechanical Code
International Plumbing Code

National Fire Protection Association (NFPA)
Life Safety Code (NFPA 101)
National Electrical Code (NFPA 70)

National Research Council of Canada
National Building Code of Canada
National Plumbing Code of Canada
National Fire Code of Canada

American Institute of Architects
AIA Documents - 2007

6. Official NCARB reference materials matrix

Per NCARB, all candidates should become familiar with the latest version of the following:

Reference	PcM	PjM	PA	PPD	PDD	CE
ADA Standards for Accessible Design U.S. Department of Justice, Latest Edition			■	■	■	
Code of Ethics and Professional Conduct AIA Office of General Counsel. The American Institute of Architects, latest edition	■					
The Architect's Handbook of Professional Practice The American Institute of Architects John Wiley & Sons, latest edition	■	■	■	■	■	■
The Architect's Studio Companion: Rules of Thumb for Preliminary Design Edward Allen and Joseph Iano John Wiley & Sons, 6th edition, 2017			■	■	■	
Architectural Acoustics. M. David Egan. J. Ross Publishing. Reprint. Original publication McGraw Hill, latest edition				■	■	
Architectural Graphic Standards The American Institute of Architects John Wiley & Sons, latest edition			■	■	■	
Building Codes Illustrated: A Guide to Understanding the International Building Code. Francis D.K. Ching and Steven R. Winkel, FAIA, PE. John Wiley & Sons, latest edition			■	■	■	
Building Construction Illustrated Francis D. K. Ching John Wiley & Sons, latest edition				■	■	
Building Structures James Ambrose and Patrick Tripeny John Wiley & Sons, 3rd edition, Latest Edition				■	■	
CSI MasterFormat. The Construction Specifications Institute, latest edition					■	■
Daylighting Handbook I Christoph Reinhart Building Technology Press, latest edition			■	■		
Dictionary of Architecture and Construction. Cyril M. Harris. McGraw-Hill, Latest edition			■	■	■	
Framework for Design Excellence American Institute of Architects Available Online			■	■		

Reference	PcM	PjM	PA	PPD	PDD	CE
Fundamentals of Building Construction: Materials and Methods Edward Allen and Joseph Iano John Wiley & Sons, latest edition						
Green Building Illustrated Francis D.K. Ching and Ian M. Shapiro Wiley, latest edition						
The Green Studio Handbook: Environmental Strategies for Schematic Design Alison G. Kwok and Walter Grondzik Routledge, latest edition						
Heating, Cooling, Lighting: Sustainable Design Methods for Architects. Norbert Lechner. John Wiley & Sons, latest edition						
The HOK Guidebook to Sustainable Design Sandra F. Mendler, William Odell, and Mary Ann Lazarus John Wiley & Sons, latest edition						
ICC A117.1-2009 Accessible and Usable Buildings and Facilities International Code Council, 2010						
International Building Code International Code Council, latest edition						
Law for Architects: What You Need to Know. Robert F. Herrmann and the Attorneys at Menaker & Herrmann LLP. W. W. Norton, latest edition						
Legislative Guidelines and Model Law/Model Regulations National Council of Architectural Registration Boards, latest edition						
Mechanical & Electrical Equipment for Buildings. Walter T. Grondzik, Alison G. Kwok, Benjamin Stein, and John S. Reynolds, Editors. John Wiley & Sons, latest edition						
Mechanical and Electrical Systems in Buildings. Richard R. Janis and William K. Y. Tao. Prentice Hall, latest edition						
Model Rules of Conduct National Council of Architectural Registration Boards, latest edition						

Reference	PcM	PjM	PA	PPD	PDD	CE
Olin's Construction Principles, Materials, and Methods. H. Leslie Simmons. John Wiley & Sons, latest edition						
Planning and Urban Design Standards American Planning Association John Wiley & Sons, latest edition						
Plumbing, Electricity, Acoustics: Sustainable Design Methods for Architecture. Norbert Lechner. John Wiley & Sons, latest edition						
Problem Seeking: An Architectural Programming Primer William M. Peña and Steven A. Parshall John Wiley & Sons, latest edition						
Professional Practice: A Guide to Turning Designs into Buildings. Paul Segal, FAIA. W. W. Norton, latest edition						
The Professional Practice of Architectural Working Drawings. Osamu A. Wakita, Nagy R. Bakhoum, and Richard M. Linde. John Wiley & Sons, latest edition						
The Project Resource Manual: CSI Manual of Practice. The Construction Specifications Institute. McGraw-Hill, latest edition						
Simplified Engineering for Architects and Builders James Ambrose and Patrick Tripeny John Wiley & Sons, latest edition						
Site Planning and Design Handbook Thomas H. Russ McGraw-Hill, latest edition						
Space Planning Basics Mark Karlen and Rob Fleming John Wiley & Sons, latest edition						
Steel Construction Manual American Institute of Steel Construction Ingram, latest edition						
Structural Design: A Practical Guide for Architects James R. Underwood and Michele Chiuini John Wiley & Sons, latest edition						
Structures Daniel Schodek and Martin Bechthold Pearson/Prentice Hall, latest edition						

Reference	PcM	PjM	PA	PPD	PDD	CE
Sun, Wind, and Light: Architectural Design Strategies G.Z. Brown and Mark DeKay John Wiley & Sons, latest edition			■	■		
Sustainable Construction: Green Building Design and Delivery Charles J. Kibert. John Wiley & Sons, latest edition				■		
A Visual Dictionary of Architecture Francis D.K. Ching John Wiley & Sons, latest edition				■	■	

The following AIA Contract Documents have content covered in the of ARE 5.0 exams. Candidates can access them for free through their NCARB Record.

Document	PcM	PjM	PA	PPD	PDD	CE
A101-2017, Standard Form of Agreement Between Owner and Contractor where the basis of payment is a Stipulated Sum		■				■
A133-2019, Standard Form of Agreement Between Owner and Construction Manager as Constructor where the basis of payment is the Cost of the Work Plus a Fee with a Guaranteed Maximum Price		■				
A195-2008, Standard Form of Agreement Between Owner and Contractor for Integrated Project Delivery		■				
A201-2017, General Conditions of the Contract for Construction		■				■
A295-2008, General Conditions of the Contract for Integrated Project Delivery		■				
A305-1986, Contractor's Qualification Statement						■
A701-2018, Instructions to Bidders						■
B101-2017, Standard Form of Agreement Between Owner and Architect	■	■				■
B195-2008, Standard Form of Agreement Between Owner and Architect for Integrated Project Delivery		■				
C401-2017, Standard Form of Agreement Between Architect and Consultant	■	■				■

Document	PcM	PjM	PA	PPD	PDD	CE
G701-2017, Change Order						
G702-1992, Application and Certificate for Payment						
G703-1992, Continuation Sheet						
G704-2017, Certificate of Substantial Completion						

The following are some extra study materials if you have some additional time and want to learn more. If you are tight on time, you can simply look through them and focus on the sections that cover your weakness:

ACI Code 318-05 (Building Code Requirements for Reinforced Concrete)
American Concrete Institute, 2005

OR
CAN/CSA-A23.1-94 (Concrete Materials and Methods of Concrete Construction) and CAN/CSA-A23.3-94 (Design of Concrete Structures for Buildings)
Canadian Standards Association

Design Value for Wood Construction
American Wood Council, 2005

Elementary Structures for Architects and Builders, Fourth Edition
Ronald E. Shaeffer
Prentice Hall, 2006

Introduction to Wood Design
Canadian Wood Council, 2005

Manual of Steel Construction: Allowable Stress Design; 9th Edition.
American Institute of Steel Construction, Inc. Chicago, Illinois, 1989

National Building Code of Canada, 2005
Parts 1, 3, 4, 9; Appendix A
Supplement
Chapters 1, 2, 4; Commentaries A, D, F, H, I

NEHRP (National Earthquake Hazards Reduction Program) Recommended Provisions for Seismic Regulations for New Buildings and Other Structures Parts 1 and 2
FEMA 2003

Simplified Building Design for Wind and Earthquake Forces
James Ambrose and Dimitry Vergun
John Wiley & Sons, 1997

Simplified Design of Concrete Structures,
Eighth Edition
James Ambrose, Patrick Tripeny
John Wiley & Sons, 2007

Simplified Design of Masonry Structures
James Ambrose
John Wiley & Sons, 1997

Simplified Design of Steel Structures, Eighth Edition
James Ambrose, Patrick Tripeny
John Wiley & Sons, 2007

Simplified Design of Wood Structures, Fifth Edition
James Ambrose
John Wiley & Sons, 2009

Simplified Mechanics and Strength of Materials, Fifth Edition
Harry Parker and James Ambrose
John Wiley & Sons, 2002

Standard Specifications Load Tables &Weight Tables for Steel Joists and Joist Girders
Steel Joist Institute, latest edition

Steel Construction Manual, Latest edition
American Institute of Steel Construction, 2006

OR
Handbook of Steel Construction, Latest edition; and *CAN/CSA-S16-01 and CISC Commentary*
Canadian Institute of Steel Construction

Steel Deck Institute Tables
Steel Deck Institute

OR
LSD Steel Deck Tables
Caradon Metal Building Products

Structural Concepts and Systems for Architects and Engineers, Second Edition
T.Y. Lin and Sidney D. Stotesbury
Van Nostrand Reinhold, 1988

Structural Design: A Practical Guide for Architects
James Underwood and Michele Chiuini
John Wiley & Sons, latest edition

Structure in Architecture: The Building of Buildings
Mario Salvadori with Robert Heller
Prentice-Hall, 1986

Understanding Structures
Fuller Moore
McGraw-Hill, 1999

Wood Design Manual and *CAN/CSA-086.1-94 and Commentary*
Canadian Wood Council

C. Other reference materials

Chen, Gang. *Building Construction: Project Management, Construction Administration, Drawings, Specs, Detailing Tips, Schedules, Checklists, and Secrets Others Don't Tell You (Architectural Practice Simplified, 2nd edition).* ArchiteG, Inc., A good introduction to the architectural practice and construction documents and service, including discussions of MasterSpec format and specification sections.

Chen, Gang. ***LEED v4 Green Associate Exam Guide (LEED GA):*** *Comprehensive Study Materials, Sample Questions, Mock Exam, Green Building LEED Certification, and Sustainability*, Book 2, LEED Exam Guide series, ArchiteG.com, the latest edition. ArchiteG, Inc. Latest Edition. This is a very comprehensive and concise book on the LEED Green Associate Exam. Some readers have passed the LEED Green Associate Exam by studying this book for 10 hours.

Ching, Francis. *Architecture: Form, Space, & Order.* Wiley, latest edition. It is one of the best architectural books that you can have. I still flip through it every now and then. It is a great book for inspiration.

Frampton, Kenneth. *Modern Architecture: A Critical History.* Thames and Hudson, London, latest edition. A valuable resource for architectural history.

Jarzombek, Mark M. (Author), Vikramaditya Prakash (Author), Francis D. K. Ching (Editor). *A Global History of Architecture.* Wiley, latest edition. A valuable and comprehensive resource for architectural history with 1000 b & w photos, 50 color photos, and 1500 b & w illustrations. It doesn't limit the topic on a Western perspective, but rather through a global vision.

Trachtenberg, Marvin and Isabelle Hyman. *Architecture: From Pre-history to Post-Modernism.* Prentice Hall, Englewood Cliffs, NJ latest edition. A valuable and comprehensive resource for architectural history.

D. Some Important Information about Architects and the Profession of Architecture

What Architects Do?

Architects plan and design houses, factories, office buildings, and other structures.

Duties

Architects typically do the following:

- meet with clients to determine objectives and requirements for structures
- give preliminary estimates on cost and construction time
- prepare structure specifications
- direct workers who prepare drawings and documents
- prepare scaled drawings, either with computer software or by hand
- prepare contract documents for building contractors
- manage construction contracts
- visit worksites to ensure that construction adheres to architectural plans
- seek new work by marketing and giving presentations

People need places to live, work, play, learn, shop, and eat. Architects are responsible for designing these places. They work on public or private projects and design both indoor and outdoor spaces. Architects can be commissioned to design anything from a single room to an entire complex of buildings.

Architects discuss the objectives, requirements, and budget of a project with clients. In some cases, architects provide various predesign services, such as feasibility and environmental impact studies, site selection, cost analyses, and design requirements.

Architects develop final construction plans after discussing and agreeing on the initial proposal with clients. These plans show the building's appearance and details of its construction. Accompanying these plans are drawings of the structural system; air-conditioning, heating, and ventilating systems; electrical systems; communications systems; and plumbing. Sometimes, landscape plans are included as well. In developing designs, architects must follow state and local building codes, zoning laws, fire regulations, and other ordinances, such as those requiring easy access to buildings for people who are disabled.

Computer-aided design and drafting (CADD) and building information modeling (BIM) have replaced traditional drafting paper and pencil as the most common methods for creating designs and construction drawings. However, hand-drawing skills are still required, especially during the conceptual stages of a project and when an architect is at a construction site.

As construction continues, architects may visit building sites to ensure that contractors follow the design, adhere to the schedule, use the specified materials, and meet work-quality standards. The job is not complete until all construction is finished, required tests are conducted, and construction costs are paid.

Architects may also help clients get construction bids, select contractors, and negotiate construction contracts.

Architects often collaborate with workers in related occupations, such as civil engineers, urban and regional planners, drafters, interior designers, and landscape architects.

Work Environment

Although architects usually work in an office, they must also travel to construction sites.

Architects held about 128,800 jobs in 2016. The largest employers of architects were as follows:

Architectural, engineering, and related services	68%
Self-employed workers	20%
Government	3%
Construction	2%

Architects spend much of their time in offices, where they meet with clients, develop reports and drawings, and work with other architects and engineers. They also visit construction sites to ensure clients' objectives are met and to review the progress of projects. Some architects work from home offices.

Work Schedules

Most architects work full time and many work additional hours, especially when facing deadlines. Self-employed architects may have more flexible work schedules.

How to Become an Architect

There are typically three main steps to becoming a licensed architect: completing a professional degree in architecture, gaining relevant experience through a paid internship, and passing the Architect Registration Examination.

Education

In all states, earning a professional degree in architecture is typically the first step to becoming an architect. Most architects earn their professional degree through a five-year Bachelor of Architecture degree program, intended for students with no previous architectural training. Many earn a master's degree in architecture, which can take one to five years in addition to the time spent earning a bachelor's degree. The amount of time required depends on the extent of the student's previous education and training in architecture.

A typical bachelor's degree program includes courses in architectural history and theory, building design with an emphasis on computer-aided design and drafting (CADD), structures, construction methods, professional practices, math, physical sciences, and liberal arts. Central to most architectural programs is the design studio, where students apply the skills and concepts learned in the classroom to create drawings and three-dimensional models of their designs.

Currently, thirty-four states require that architects hold a professional degree in architecture from one of the 123 schools of architecture accredited by the National Architectural Accrediting Board (NAAB). State licensing requirements can be found at the National Council of Architectural

Registration Boards (NCARB). In the states that do not have that requirement, applicants can become licensed with eight to thirteen years of related work experience in addition to a high school diploma. However, most architects in these states still obtain a professional degree in architecture.

Training

All state architectural registration boards require architecture graduates to complete a lengthy paid internship—generally three years of experience—before they may sit for the Architect Registration Examination. Most new graduates complete their training period by working at architectural firms through the Architectural Experience Program (AXP), a program run by NCARB that guides students through the internship process. Some states allow a portion of the training to occur in the offices of employers in related careers, such as engineers and general contractors. Architecture students who complete internships while still in school can count some of that time toward the three-year training period.

Interns in architectural firms may help design part of a project. They may help prepare architectural documents and drawings, build models, and prepare construction drawings on CADD. Interns may also research building codes and write specifications for building materials, installation criteria, the quality of finishes, and other related details. Licensed architects will take the documents that interns produce, make edits to them, finalize plans, and then sign and seal the documents.

Licenses, Certifications, and Registrations

All states and the District of Columbia require architects to be licensed. Licensing requirements typically include completing a professional degree in architecture, gaining relevant experience through a paid internship, and passing the Architect Registration Examination.

Most states also require some form of continuing education to keep a license, and some additional states are expected to adopt mandatory continuing education. Requirements vary by state but usually involve additional education through workshops, university classes, conferences, self-study courses, or other sources.

A growing number of architects voluntarily seek certification from NCARB. This certification makes it easier to become licensed in other states, because it is the primary requirement for reciprocity of licensing among state boards that are NCARB members. In 2014, approximately one-third of all licensed architects had the certification.

Advancement

After many years of work experience, some architects advance to become architectural and engineering managers. These managers typically coordinate the activities of employees and may work on larger construction projects.

Important Qualities

Analytical skills. Architects must understand the content of designs and the context in which they were created. For example, architects must understand the locations of mechanical systems and how those systems affect building operations.

Communication skills. Architects share their ideas, both in oral presentations and in writing, with clients, other architects, and workers who help prepare drawings. Many also give presentations to explain their ideas and designs.

Creativity. Architects design the overall look of houses, buildings, and other structures. Therefore, the final product should be attractive and functional.

Organizational skills. Architects often manage contracts. Therefore, they must keep records related to the details of a project, including total cost, materials used, and progress.

Technical skills. Architects need to use CADD technology to create plans as part of building information modeling (BIM).

Visualization skills. Architects must be able to see how the parts of a structure relate to each other. They also must be able to visualize how the overall building will look once completed.

Pay

The median annual wage for architects was $79,380 in May 2018. The median wage is the wage at which half the workers in an occupation earned more than that amount and half earned less. The lowest 10 percent earned less than $48,020, and the highest ten percent earned more than $138,120.

In May 2018, the median annual wages for architects in the top industries in which they worked were as follows:

Government	$92,940
Architectural, engineering, and related services	$78,460
Construction	$78,110

Most architects work full time and many work additional hours, especially when facing deadlines. Self-employed architects may have more flexible work hours.

Job Outlook

Employment of architects is projected to grow 4 percent from 2016 to 2026, slower than the average for all occupations.

Architects will be needed to make plans and designs for the construction and renovation of homes, offices, retail stores, and other structures. Many school districts and universities are expected to build new facilities or renovate existing ones. In addition, demand is expected for more healthcare facilities as the baby-boomer population ages and as more individuals use healthcare services. The construction of new retail establishments may also require more architects.

Demand is projected for architects with a knowledge of "green design," also called sustainable design. Sustainable design emphasizes the efficient use of resources, such as energy and water conservation; waste and pollution reduction; and environmentally friendly design, specifications, and materials. Rising energy costs and increased concern about the environment have led to many new buildings being built with more sustainable designs.

The use of CADD and, more recently, BIM, has made architects more productive. These technologies have allowed architects to do more work without the help of drafters while making it easier to share the work with engineers, contractors, and clients.

Job Prospects
With a high number of students graduating with degrees in architecture, very strong competition for internships and jobs is expected. Competition for jobs will be especially strong at the most prestigious architectural firms. Those with up-to-date technical skills—including a strong grasp of CADD and BIM—and experience in sustainable design will have an advantage.

Employment of architects is strongly tied to the activity of the construction industry. Therefore, these workers may experience periods of unemployment when there is a slowdown in requests for new projects or when the overall level of construction falls.

State & Area Data
Occupational Employment Statistics (OES)
The Occupational Employment Statistics (OES) program produces employment and wage estimates annually for over 800 occupations. These estimates are available for the nation as a whole, for individual states, and for metropolitan and nonmetropolitan areas. The link below goes to OES data maps for employment and wages by state and area.
https://www.bls.gov/oes/current/oes171011.htm#st

Projections Central
Occupational employment projections are developed for all states by Labor Market Information (LMI) or individual state Employment Projections offices. All state projections data are available at www.projectionscentral.com. Information on this site allows projected employment growth for an occupation to be compared among states or to be compared within one state. In addition, states may produce projections for areas; there are links to each state's websites where these data may be retrieved.

CareerOneStop
CareerOneStop includes hundreds of occupational profiles with data available by state and metro area. There are links in the left-hand side menu to compare occupational employment by state and occupational wages by local area or metro area. There is also a salary info tool to search for wages by zip code.

Related Occupations
Architects design buildings and related structures. Construction managers, like architects, also plan and coordinate activities concerned with the construction and maintenance of buildings and facilities. Others who engage in similar work are landscape architects, civil engineers, urban and regional planners, and designers, including interior designers, commercial and industrial designers, and graphic designers.

Sources of Additional Information

Disclaimer:
Links to non-BLS Internet sites are provided for your convenience and do not constitute an endorsement.

Information about education and careers in architecture can be obtained from:

- The American Institute of Architects, 1735 New York Ave. NW., Washington, DC 20006. Internet: http://www.aia.org
- National Architectural Accrediting Board: http://www.naab.org/
- National Council of Architectural Registration Boards, Suite 1100K, 1801 K St. NW., Washington, D.C. 20006. Internet: http://www.ncarb.org
  OOH ONET Codes 17-1011.00"

Source: Bureau of Labor Statistics, U.S. Department of Labor, *Occupational Outlook Handbook*, Architects, on the Internet at https://www.bls.gov/ooh/architecture-and-engineering/architects.htm (visited June 06, 2019).

Last Modified Date: Friday, April 12, 2019

Note:
Please check the website above for the latest information.

E. AIA Compensation Survey

Every 3 years, AIA publishes a Compensation Survey for various positions at architectural firms across the country. It is a good idea to find out the salary before you make the final decision to become an architect. If you are already an architect, it is also a good idea to determine if you are underpaid or overpaid.

See following link for some sample pages for the latest AIA Compensation Survey:

https://www.aia.org/resources/8066-aia-compensation-report

F. So … You would Like to Study Architecture

To study architecture, you need to learn how to draft, how to understand and organize spaces and the interactions between interior and exterior spaces, how to do design, and how to communicate effectively. You also need to understand the history of architecture.

As an architect, a leader for a team of various design professionals, you not only need to know architecture, but also need to understand enough of your consultants' work to be able to coordinate them. Your consultants include soils and civil engineers, landscape architects, structural, electrical, mechanical, and plumbing engineers, interior designers, sign consultants, etc.

There are two major career paths for you in architecture: practice as an architect or teach in colleges or universities. The earlier you determine which path you are going to take, the more likely you will be successful at an early age. Some famous and well-respected architects, like my USC alumnus Frank Gehry, have combined the two paths successfully. They teach at the universities and have their own architectural practice. Even as a college or university professor, people respect you more if you have actual working experience and have some built projects. If you only teach in colleges or universities but have no actual working experience and have no built projects, people will consider you as a "paper" architect, and they are not likely to take you seriously, because they will think you probably do not know how to put a real building together.

In the U.S., if you want to practice architecture, you need to obtain an architect's license. It requires a combination of passing scores on the Architectural Registration Exam (ARE) and 8 years of education and/or qualified working experience, including at least 1 year of working experience in the U.S. Your working experience needs to be under the supervision of a licensed architect to be counted as qualified working experience for your architect's license.

If you work for a landscape architect or civil engineer or structural engineer, some states' architectural licensing boards will count your experience at a discounted rate for the qualification of your architect's license. For example, 2 years of experience working for a civil engineer may be counted as 1 year of qualified experience for your architect's license. You need to contact your state's architectural licensing board for specific licensing requirements for your state.

If you want to teach in colleges or universities, you probably want to obtain a master's degree or a Ph.D. It is not very common for people in the architectural field to have a Ph.D. One reason is that there are few Ph.D. programs for architecture. Another reason is that architecture is considered a profession and requires a license. Many people think an architect's license is more important than a Ph.D. degree. In many states, you need to have an architect's license to even use the title "architect," or the terms "architectural" or "architecture" to advertise your service. You cannot call yourself an architect if you do not have an architect's license, even if you have a Ph.D. in architecture. Violation of these rules brings punishment.

To become a tenured professor, you need to have a certain number of publications and pass the evaluation for the tenure position. Publications are very important for tenure track positions. Some people say for the tenured track positions in universities and colleges, it is "publish or perish."

The American Institute of Architects (AIA) is the national organization for the architectural profession. Membership is voluntary. There are different levels of AIA membership. Only licensed architects can be (full) AIA members. If you are an architectural student or an intern but not a licensed architect yet, you can join as an associate AIA member. Contact AIA for detailed information.

The National Council of Architectural Registration Boards (NCARB) is a nonprofit federation of architectural licensing boards. It has some very useful programs, such as IDP, to assist you in obtaining your architect's license. Contact NCARB for detailed information.

Back Page Promotion

You may be interested in some other books written by Gang Chen:

A. **ARE Mock Exam series & ARE Exam Guide series. See the following link:**
http://www.GreenExamEducation.com

B. **LEED Exam Guides series. See the following link:**
http://www.GreenExamEducation.com

C. ***Building Construction:*** *Project Management, Construction Administration, Drawings, Specs, Detailing Tips, Schedules, Checklists, and Secrets Others Don't Tell You (Architectural Practice Simplified, 2nd edition)*
http://www.GreenExamEducation.com

D. ***Planting Design Illustrated***
http://www.GreenExamEducation.com

ARE Mock Exam Series & ARE Exam Guide Series

ARE 5.0 Mock Exam Series

Practice Management (PcM) ARE 5.0 Mock Exam (Architect Registration Examination): *ARE 5.0 Overview, Exam Prep Tips, Hotspots, Case Studies, Drag-and-Place, Solutions and Explanations.* **ISBN**: 9781612650388

Project Management (PjM) ARE 5.0 Mock Exam (Architect Registration Examination): *ARE 5.0 Overview, Exam Prep Tips, Hotspots, Case Studies, Drag-and-Place, Solutions and Explanations.* **ISBN**: 9781612650371

Programming & Analysis (PA) ARE 5.0 Mock Exam (Architect Registration Exam): *ARE 5.0 Overview, Exam Prep Tips, Hotspots, Case Studies, Drag-and-Place, Solutions and Explanations.* **ISBN**: 9781612650326

Project Planning & Design (PPD) ARE 5.0 Mock Exam (Architect Registration Examination): *ARE 5.0 Overview, Exam Prep Tips, Hotspots, Case Studies, Drag-and-Place, Solutions and Explanations.* **ISBN**: 9781612650296

Project Development & Documentation (PDD) ARE 5.0 Mock Exam (Architect Registration Examination): *ARE 5.0 Overview, Exam Prep Tips, Hotspots, Case Studies, Drag-and-Place, Solutions and Explanations*
ISBN: 9781612650258

Construction & Evaluation (CE) ARE 5.0 Mock Exam (Architect Registration Examination): *ARE 5.0 Overview, Exam Prep Tips, Hotspots, Case Studies, Drag-and-Place, Solutions and Explanations*
ISBN: 9781612650241

Mock California Supplemental Exam (CSE of Architect Registration Examination): *CSE Overview, Exam Prep Tips, General Section and Project Scenario Section, Questions, Solutions and Explanations.* **ISBN**: 9781612650159

ARE 5.0 Exam Guide Series

Practice Management (PcM) ARE 5.0 Exam Guide (Architect Registration Examination): *ARE 5.0 Overview, Exam Prep Tips, Guide, and Critical Content.* **ISBN**: 9781612650333

Project Management (PjM) ARE 5.0 Exam Guide (Architect Registration Examination): *ARE 5.0 Overview, Exam Prep Tips, Guide, and Critical Content.* **ISBN**: 9781612650418

Programming & Analysis (PA) ARE 5.0 Exam Guide (Architect Registration Examination): *ARE 5.0 Overview, Exam Prep Tips, Guide, and Critical Content.*
ISBN: 9781612650487

Construction and Evaluation (CE) ARE 5 Exam Guide (Architect Registration Exam):
ARE 5.0 Overview, Exam Prep Tips, Guide, and Critical Content
ISBN: 9781612650432

Other books in the ARE 5.0 Exam Guide Series are being produced. Our goal is to produce one mock exam book *plus* one guidebook for each of the ARE 5.0 exam divisions. See the following link for the latest information:
http://www.GreenExamEducation.com

LEED Exam Guides series: Comprehensive Study Materials, Sample Questions, Mock Exam, Building LEED Certification and Going Green

LEED (Leadership in Energy and Environmental Design) is the most important trend of development, and it is revolutionizing the construction industry. It has gained tremendous momentum and has a profound impact on our environment.

From LEED Exam Guides series, you will learn how to

1. Pass the LEED Green Associate Exam and various LEED AP + exams (each book will help you with a specific LEED exam).

2. Register and certify a building for LEED certification.

3. Understand the intent for each LEED prerequisite and credit.

4. Calculate points for a LEED credit.

5. Identify the responsible party for each prerequisite and credit.

6. Earn extra credit (exemplary performance) for LEED.

7. Implement the local codes and building standards for prerequisites and credit.

8. Receive points for categories not yet clearly defined by USGBC.

There is currently NO official book on the LEED Green Associate Exam, and most of the existing books on LEED and LEED AP are too expensive and too complicated to be practical and helpful. The pocket guides in LEED Exam Guides series fill in the blanks, demystify LEED, and uncover the tips, codes, and jargon for LEED as well as the true meaning of "going green." They will set up a solid foundation and fundamental framework of LEED for you. Each book in the LEED Exam Guides series covers every aspect of one or more specific LEED rating system(s) in plain and concise language and makes this information understandable to all people.

These pocket guides are small and easy to carry around. You can read them whenever you have a few extra minutes. They are indispensable books for all people—administrators; developers; contractors; architects; landscape architects; civil, mechanical, electrical, and plumbing engineers; interns; drafters; designers; and other design professionals.

Why is the LEED Exam Guides series needed?

A number of books are available that you can use to prepare for the LEED exams:

1. *USGBC Reference Guides*. You need to select the correct version of the *Reference Guide* for your exam.

   The *USGBC Reference Guides* are comprehensive, but they give too much information. For example, *The LEED 2009 Reference Guide for Green Building Design and Construction (BD&C)* has about 700 oversized pages. Many of the calculations in the books are too detailed for the exam. They are also expensive (approximately $200 each, so most people may not buy them for their personal use, but instead, will seek to share an office copy).

   It is good to read a reference guide from cover to cover if you have the time. The problem is not too many people have time to read the whole reference guide. Even if you do read the whole guide, you may not remember the important issues to pass the LEED exam. You need to reread the material several times before you can remember much of it.

   Reading the reference guide from cover to cover without a guidebook is a difficult and inefficient way of preparing for the LEED AP Exam, because you do NOT know what USGBC and GBCI are looking for in the exam.

2. The USGBC workshops and related handouts are concise, but they do not cover extra credits (exemplary performance). The workshops are expensive, costing approximately $450 each.

3. Various books published by a third party are available on Amazon, bn.com and books.google.com. However, most of them are not very helpful.

   There are many books on LEED, but not all are useful.

   LEED Exam Guides series will fill in the blanks and become a valuable, reliable source:

   a. They will give you more information for your money. Each of the books in the LEED Exam Guides series has more information than the related USGBC workshops.

   b. They are exam-oriented and more effective than the USGBC reference guides.

   c. They are better than most, if not all, of the other third-party books. They give you comprehensive study materials, sample questions and answers, mock exams and answers, and critical information on building LEED certification and going green. Other third-party books only give you a fraction of the information.

   d. They are comprehensive yet concise. They are small and easy to carry around. You can read them whenever you have a few extra minutes.

   e. They are great timesavers. I have highlighted the important information that you need to understand and MEMORIZE. I also make some acronyms and short sentences to help you easily remember the credit names.

It should take you about 1 or 2 weeks of full-time study to pass each of the LEED exams. I have met people who have spent 40 hours to study and passed the exams.

You can find sample texts and other information on the LEED Exam Guides series in customer discussion sections under each of my book's listing on Amazon, bn.com and books.google.com.

What others are saying about *LEED GA Exam Guide* (Book 2, LEED Exam Guide series):

"Finally! A comprehensive study tool for LEED GA Prep!

"I took the 1-day Green LEED GA course and walked away with a power point binder printed in very small print—which was missing MUCH of the required information (although I didn't know it at the time). I studied my little heart out and took the test, only to fail it by 1 point. Turns out I did NOT study all the material I needed to in order to pass the test. I found this book, read it, marked it up, retook the test, and passed it with a 95%. Look, we all know the LEED GA exam is new and the resources for study are VERY limited. This one is the VERY best out there right now. I highly recommend it."
—ConsultantVA

"Complete overview for the LEED GA exam

"I studied this book for about 3 days and passed the exam ... if you are truly interested in learning about the LEED system and green building design, this is a great place to start."
—K.A. Evans

"A Wonderful Guide for the LEED GA Exam

"After deciding to take the LEED Green Associate exam, I started to look for the best possible study materials and resources. From what I thought would be a relatively easy task, it turned into a tedious endeavor. I realized that there are vast amounts of third-party guides and handbooks. Since the official sites offer little to no help, it became clear to me that my best chance to succeed and pass this exam would be to find the most comprehensive study guide that would not only teach me the topics, but would also give me a great background and understanding of what LEED actually is. Once I stumbled upon Mr. Chen's book, all my needs were answered. This is a great study guide that will give the reader the most complete view of the LEED exam and all that it entails.

"The book is written in an easy-to-understand language and brings up great examples, tying the material to the real world. The information is presented in a coherent and logical way, which optimizes the learning process and does not go into details that will not be needed for the LEED Green Associate Exam, as many other guides do. This book stays dead on topic and keeps the reader interested in the material.

"I highly recommend this book to anyone that is considering the LEED Green Associate Exam. I learned a great deal from this guide, and I am feeling very confident about my chances for passing my upcoming exam."
—Pavel Geystrin

"Easy to read, easy to understand

"I have read through the book once and found it to be the perfect study guide for me. The author does a great job of helping you get into the right frame of mind for the content of the exam. I had started by studying the Green Building Design and Construction reference guide for LEED projects produced by the USGBC. That was the wrong approach, simply too much information with very little retention. At 636 pages in textbook format, it would have been a daunting task to get through it. Gang Chen breaks down the points, helping to minimize the amount of information but maximizing the content I was able to absorb. I plan on going through the book a few more times, and I now believe I have the right information to pass the LEED Green Associate Exam."
—Brian Hochstein

"All in one—LEED GA prep material

"Since the LEED Green Associate exam is a newer addition by USGBC, there is not much information regarding study material for this exam. When I started looking around for material, I got really confused about what material I should buy. This LEED GA guide by Gang Chen is an answer to all my worries! It is a very precise book with lots of information, like how to approach the exam, what to study and what to skip, links to online material, and tips and tricks for passing the exam. It is like the 'one stop shop' for the LEED Green Associate Exam. I think this book can also be a good reference guide for green building professionals. A must-have!"
—SwatiD

"An ESSENTIAL LEED GA Exam Reference Guide

"This book is an invaluable tool in preparation for the LEED Green Associate (GA) Exam. As a practicing professional in the consulting realm, I found this book to be all-inclusive of the preparatory material needed for sitting the exam. The information provides clarity to the fundamental and advanced concepts of what LEED aims to achieve. A tremendous benefit is the connectivity of the concepts with real-world applications.

"The author, Gang Chen, provides a vast amount of knowledge in a very clear, concise, and logical media. For those that have not picked up a textbook in a while, it is very manageable to extract the needed information from this book. If you are taking the exam, do yourself a favor and purchase a copy of this great guide. Applicable fields: Civil Engineering, Architectural Design, MEP, and General Land Development."
—Edwin L. Tamang

Note:
*Other books in the **LEED Exam Guides series** are published or in the process of being produced. At least **one book will eventually be produced for each of the LEED exams.** The series include:*

LEED v4 Green Associate Exam Guide (LEED GA): *Comprehensive Study Materials, Sample Questions, Mock Exam, Green Building LEED Certification, and Sustainability*, LEED Exam Guide series, ArchiteG.com. Latest Edition.

LEED GA MOCK EXAMS (LEED v4): *Questions, Answers, and Explanations: A Must-Have for the LEED Green Associate Exam, Green Building LEED Certification, and Sustainability*, LEED Exam Guide series, ArchiteG.com. Latest Edition

LEED v4 BD&C EXAM GUIDE: *A Must-Have for the LEED AP BD+C Exam: Comprehensive Study Materials, Sample Questions, Mock Exam, Green Building Design and Construction, LEED Certification, and Sustainability*, LEED Exam Guide series, ArchiteG.com. Latest Edition.

LEED v4 BD&C MOCK EXAMS: *Questions, Answers, and Explanations: A Must-Have for the LEED AP BD+C Exam, Green Building LEED Certification, and Sustainability*, LEED Exam Guide series, ArchiteG.com. Latest Edition.

LEED v4 ID&C Exam Guide: *A Must-Have for the LEED AP ID+C Exam: Study Materials, Sample Questions, Green Interior Design and Construction, Green Building LEED Certification, and Sustainability*, LEED Exam Guide series, ArchiteG.com. Latest Edition.

LEED v4 AP ID+C MOCK EXAM: *Questions, Answers, and Explanations: A Must-Have for the LEED AP ID+C Exam, Green Building LEED Certification, and Sustainability*. LEED Exam Guide series, ArchiteG.com. Latest Edition.

LEED v4 AP O+M MOCK EXAM: Questions, Answers, and Explanations: *A Must-Have for the LEED AP O+M Exam, Green Building LEED Certification, and Sustainability*. LEED Exam Guide series, ArchiteG.com. Latest Edition.

LEED v4 O&M EXAM GUIDE: *A Must-Have for the LEED AP O+M Exam: Comprehensive Study Materials, Sample Questions, Mock Exam, Green Building Operations and Maintenance, LEED Certification, and Sustainability*, LEED Exam Guide series, ArchiteG.com. Latest Edition.

LEED v4 HOMES EXAM GUIDE: *A Must-Have for the LEED AP Homes Exam: Comprehensive Study Materials, Sample Questions, Mock Exam, Green Building LEED Certification, and Sustainability*, LEED Exam Guide series, ArchiteG.com. Latest Edition.

LEED v4 ND EXAM GUIDE: *A Must-Have for the LEED AP Neighborhood Development Exam: Comprehensive Study Materials, Sample Questions, Mock Exam, Green Building LEED Certification, and Sustainability*, LEED Exam Guide series, ArchiteG.com. Latest Edition.

How to order these books:
You can order the books listed above at:
http://www.GreenExamEducation.com

OR
http://www.ArchiteG.com

Building Construction

Project Management, Construction Administration, Drawings, Specs, Detailing Tips, Schedules, Checklists, and Secrets Others Don't Tell You (Architectural Practice Simplified, 2nd edition)

Learn the Tips, Become One of Those Who Know Building Construction and Architectural Practice, and Thrive!

For architectural practice and building design and construction industry, there are two kinds of people: those who know, and those who don't. The tips of building design and construction and project management have been undercover—until now.

Most of the existing books on building construction and architectural practice are too expensive, too complicated, and too long to be practical and helpful. This book simplifies the process to make it easier to understand and uncovers the tips of building design and construction and project management. It sets up a solid foundation and fundamental framework for this field. It covers every aspect of building construction and architectural practice in plain and concise language and introduces it to all people. Through practical case studies, it demonstrates the efficient and proper ways to handle various issues and problems in architectural practice and building design and construction industry.

It is for ordinary people and aspiring young architects as well as seasoned professionals in the construction industry. For ordinary people, it uncovers the tips of building construction; for aspiring architects, it works as a construction industry survival guide and a guidebook to shorten the process in mastering architectural practice and climbing up the professional ladder; for seasoned architects, it has many checklists to refresh their memory. It is an indispensable reference book for ordinary people, architectural students, interns, drafters, designers, seasoned architects, engineers, construction administrators, superintendents, construction managers, contractors, and developers.

You will learn:

1. How to develop your business and work with your client.
2. The entire process of building design and construction, including programming, entitlement, schematic design, design development, construction documents, bidding, and construction administration.
3. How to coordinate with governing agencies, including a county's health department and a city's planning, building, fire, public works departments, etc.
4. How to coordinate with your consultants, including soils, civil, structural, electrical, mechanical, plumbing engineers, landscape architects, etc.
5. How to create and use your own checklists to do quality control of your construction documents.
6. How to use various logs (i.e., RFI log, submittal log, field visit log, etc.) and lists (contact list, document control list, distribution list, etc.) to organize and simplify your work.
7. How to respond to RFI, issue CCDs, review change orders, submittals, etc.
8. How to make your architectural practice a profitable and successful business.

Planting Design Illustrated

A Must-Have for Landscape Architecture: A Holistic Garden Design Guide with Architectural and Horticultural Insight, and Ideas from Famous Gardens in Major Civilizations

One of the most significant books on landscaping!

This is one of the most comprehensive books on planting design. It fills in the blanks of the field and introduces poetry, painting, and symbolism into planting design. It covers in detail the two major systems of planting design: formal planting design and naturalistic planting design. It has numerous line drawings and photos to illustrate the planting design concepts and principles. Through in-depth discussions of historical precedents and practical case studies, it uncovers the fundamental design principles and concepts, as well as the underpinning philosophy for planting design. It is an indispensable reference book for landscape architecture students, designers, architects, urban planners, and ordinary garden lovers.

What Others Are Saying about *Planting Design Illustrated* …

"I found this book to be absolutely fascinating. You will need to concentrate while reading it, but the effort will be well worth your time."
—Bobbie Schwartz, former president of APLD (Association of Professional Landscape Designers) and author of *The Design Puzzle: Putting the Pieces Together*.

"This is a book that you have to read, and it is more than well worth your time. Gang Chen takes you well beyond what you will learn in other books about basic principles like color, texture, and mass."
—Jane Berger, editor & publisher of gardendesignonline

"As a longtime consumer of gardening books, I am impressed with Gang Chen's inclusion of new information on planting design theory for Chinese and Japanese gardens. Many gardening books discuss the beauty of Japanese gardens, and a few discuss the unique charms of Chinese gardens, but this one explains how Japanese and Chinese history, as well as geography and artistic traditions, bear on the development of each country's style. The material on traditional Western garden planting is thorough and inspiring, too. *Planting Design Illustrated* definitely rewards repeated reading and study. Any garden designer will read it with profit."
—Jan Whitner, editor of the *Washington Park Arboretum Bulletin*

"Enhanced with an annotated bibliography and informative appendices, *Planting Design Illustrated* offers an especially "reader friendly" and practical guide that makes it a very strongly recommended addition to personal, professional, academic, and community library gardening & landscaping reference collection and supplemental reading list."
—Midwest Book Review

"Where to start? *Planting Design Illustrated* is, above all, fascinating and refreshing! Not something the lay reader encounters every day, the book presents an unlikely topic in an easily digestible, easy-to-follow way. It is superbly organized with a comprehensive table of contents, bibliography, and appendices. The writing, though expertly informative, maintains its accessibility throughout and is a joy to read. The detailed and beautiful illustrations expanding on the concepts presented were my favorite portion. One of the finest books I've encountered in this contest in the past 5 years."
—Writer's Digest 16th Annual International Self-Published Book Awards Judge's Commentary

"The work in my view has incredible application to planting design generally and a system approach to what is a very difficult subject to teach, at least in my experience. Also featured is a very beautiful philosophy of garden design principles bordering poetry. It's my strong conviction that this work needs to see the light of day by being published for the use of professionals, students & garden enthusiasts."
—Donald C. Brinkerhoff, FASLA, chairman and CEO of Lifescapes International, Inc.

Index

Made in the USA
Middletown, DE
20 January 2022